Lecture Notes in Physics

New Series m: Monographs

Springer-Verlag Berlin Heidelberg GmbH

The Editorial Policy for Monographs

The series Lecture Notes in Physics reports new developments in physical research and teaching - quickly, informally, and at a high level. The type of material considered for publication in the New Series m includes monographs presenting original research or new angles in a classical field. The timeliness of a manuscript is more important than its form, which may be preliminary or tentative. Manuscripts should be reasonably self-contained. They will often present not only results of the author(s) but also related work by other people and will provide sufficient motivation, examples, and applications.

The manuscripts or a detailed description thereof should be submitted either to one of the series editors or to the managing editor. The proposal is then carefully refereed. A final decision concerning publication can often only be made on the basis of the complete manuscript, but otherwise the editors will try to make a preliminary decision as definite as they can on the basis of the available information.

Manuscripts should be no less than 100 and preferably no more than 400 pages in length. Final manuscripts should preferably be in English, or possibly in French or German. They should include a table of contents and an informative introduction accessible also to readers not particularly familiar with the topic treated. Authors are free to use the material in other publications. However, if extensive use is made elsewhere, the publisher should be informed. Authors receive jointly 50 complimentary copies of their book. They are entitled to purchase further copies of their book at a reduced rate. As a rule no reprints of individual contributions can be supplied. No royalty is paid on Lecture Notes in Physics volumes. Commitment to publish is made by letter of interest rather than by signing a formal contract. Springer-Verlag secures the copyright for each volume.

The Production Process

The books are hardbound, and quality paper appropriate to the needs of the author(s) is used. Publication time is about ten weeks. More than twenty years of experience guarantee authors the best possible service. To reach the goal of rapid publication at a low price the technique of photographic reproduction from a camera-ready manuscript was chosen. This process shifts the main responsibility for the technical quality considerably from the publisher to the author. We therefore urge all authors to observe very carefully our guidelines for the preparation of camera-ready manuscripts, which we will supply on request. This applies especially to the quality of figures and halftones submitted for publication. Figures should be submitted as originals or glossy prints, as very often Xerox copies are not suitable for reproduction. For the same reason, any writing within figures should not be smaller than 2.5 mm. It might be useful to look at some of the volumes already published or, especially if some atypical text is planned, to write to the Physics Editorial Department of Springer-Verlag direct. This avoids mistakes and time-consuming correspondence during the production period.

As a special service, we offer free of charge $\text{L\!A\!T\!}_{\text{E}}\text{X}$ and $\text{T\!}_{\text{E}}\text{X}$ macro packages to format the text according to Springer-Verlag's quality requirements. We strongly recommend authors to make use of this offer, as the result will be a book of considerably improved technical quality.

Manuscripts not meeting the technical standard of the series will have to be returned for improvement.

For further information please contact Springer-Verlag, Physics Editorial Department II, Tiergartenstrasse 17, D-69121 Heidelberg, Germany.

Reinhard Alkofer   Hugo Reinhardt

Chiral Quark Dynamics

Springer-Verlag Berlin Heidelberg GmbH

Authors

Reinhard Alkofer
Hugo Reinhardt
Institute for Theoretical Physics
University of Tübingen
Auf der Morgenstelle 14
D-72076 Tübingen, Germany

Cataloging-in-Publication Data applied for.

Die Deutsche Bibliothek - CIP-Einheitsaufnahme

Alkofer, Reinhard:
Chiral quark dynamics / Reinhard Alkofer ; Hugo Reinhardt. -
Berlin ; Heidelberg ; New York ; Barcelona ; Budapest ; Hong
Kong ; London ; Milan ; Paris ; Tokyo : Springer, 1995
  (Lecture notes in physics : N.s. M, Monographs ; Vol. 33)

NE: Reinhardt, Hugo:; Lecture notes in physics / M

ISBN 978-3-662-14020-8   ISBN 978-3-540-49454-6 (eBook)
DOI 10.1007/978-3-540-49454-6

© Springer-Verlag Berlin Heidelberg 1995
Originally published by Springer-Verlag Berlin Heidelberg New York in 1995
Softcover reprint of the hardcover 1st edition 1995

Typesetting: Camera-ready by authors using T_EX
SPIN: 10127391   55/3142-543210 - Printed on acid-free paper

Preface

These Lecture Notes are based partly on a lecture given by one of us (H.R.) at Tübingen University in Spring 1991 and partly on a lecture given at the Egyptian–German Spring School "Particle and Nuclear Phyics" in Cairo in April 1992. They are addressed to graduate students and young research workers in theoretical physics. Some knowledge of quantum field theory, especially on functional integral techniques, are required. These Notes are intended to give a pedagogical introduction into the description of hadrons, i.e., mesons and baryons, within a quark model based on a chirally invariant quantum field theory. A more detailed description of of the subject in Chap. 4, the chiral soliton of the Nambu–Jona-Lasinio model, is given in a recent review [AHW95].

In these Notes we have used results from recent research papers. It is a pleasure to thank our coauthors for their fruitful collaboration. We are especially indebted to Dr. Herbert Weigel who carried the main load in the investigations concerning the NJL soliton. We thank also Albrecht Buck and Udo Zückert for their valuable contributions. Furthermore we also acknowledge discussions with Kurt Langfeld, Lorenz von Smekal, Christian Weiss, Roman Friedrich, Axel Bender, Gerhard Hellstern, and Jürgen Schlienz.

Tübingen, January 1995
R. Alkofer
H. Reinhardt

Contents

Chapter 1

Introduction

According to our present understanding **Quantum Chromo Dynamics** (QCD) is the theory of strong interaction. QCD is a non–Abelian Yang–Mills theory with gauge group SU(3). The elementary fermions of the theory, the quarks, live in the fundamental representation of the gauge group and carry therefore a quantum number called **color** which can take three different values, and which may be named *e.g.* red, green and blue. The interaction between the quarks is mediated by the gauge bosons, the gluons, which are defined in the adjoint representation of SU(3). They come hence with eight different color species. Due to the non–Abelian structure of the theory the gluons couple not only to the quarks but have cubic and quartic self–interactions. This self–interaction of the gluons provides the antiscreening of color charges in QCD contrary to screening of electric charge in QED. As a consequence the running coupling constant of QCD is large at small energies or momenta and small at high energies. The gluon self–interaction is responsible for asymptotic freedom and presumably also for confinement.

In nature one did not observe colored objects. It is commonly assumed that color is confined, *i.e.* only colorless particles which are composed of colored quarks and gluons can be observed (confinement hypothesis). However, for the moment no rigorous proof of confinement within QCD exists. Besides color the quarks, but not the gluons, carry a further quantum number called **flavor**. This quantum number is manifest in the physical hadron spectrum. In fact, it is the quark flavor content which determines the physical properties of the hadrons. At present we know six quark flavors: up, down, strange, charm, bottom and top (u, d, s, c, b, t). The quark flavors differ in their current mass m^0 which is presumably of electroweak origin, $m_u^0 < m_d^0 \ll m_s^0 \ll m_c^0 \ll m_b^0 \ll m_t^0$. The difference in the current quark masses triggers via the strong interaction (gluon exchange) different effective properties of the quark flavors at low energies. These "effective" quarks are called "constituent quarks". These constituent quarks, rather than the current quarks, enter the low–energy quark models.

The dynamics of the different quark flavors mediated by the strong interaction will be the subject of these lectures: **Quark Flavor Dynamics** (QFD). The basic assumption or philosophy of QFD is: Since flavor but not color is seen in the low–energy experimental hadron spectra the strong color correlations should result at low energies in an effective color blind but flavor dependent interaction of the quarks from which the physical hadron properties should be calculable.

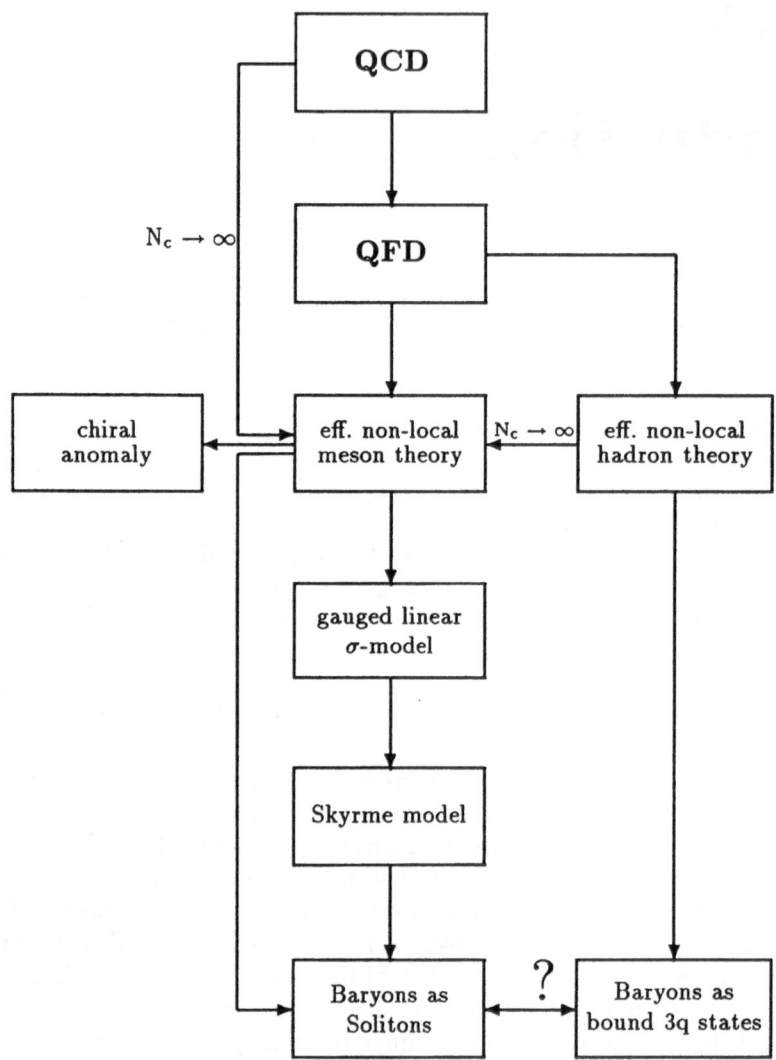

Fig. 1.1. Outline of these lecture notes.

The organization of this lecture notes is displayed in Fig. 1.1. First, we will outline how to extract the effective QFD from QCD. The basic idea is an expansion in terms of the color octet flavor singlet quark currents, see Chap. 2. Also a recently suggested approach to the non–perturbative regime of QCD, the field strength approach, leads to a qualitatively similar result. We will end up with an effective quark theory with (local) two–body interaction. Then we are going to discuss and compare the invariance properties of QCD and QFD. Special emphasis will be put on **chiral symmetry** because it plays a central role in all the following topics. Rewriting the effective quark interaction from a current–current coupling with color octet flavor singlet currents into a color singlet flavor dependent interaction is done with the help of generalized Fierz identities. Path integral bosonization is used to derive an effective non–local meson theory (Chap. 3). It is demonstrated that chiral symmetry is dynamically broken within this theory. In Sect. 3.3 we derive the Bethe–Salpeter equation for the Goldstone bosons, the pseudoscalar mesons.* With the help of a covariant gradient expansion (whose mathematical details are presented in the Appendix A) we arrive at a gauged linear σ–model. This model displays all the wisdom of the sixties and seventies about hadron physics. Besides a lot of other relations we will recover well–known low–energy theorems like PCAC, Weinberg's relation or the KSFR relation. We will also see that this effective model describes the properties of mesons reasonably well. In a next step we will neglect fluctuations in the scalar field (*i.e.* constrain the system on the chiral circle) thereby arriving at a special non–linear σ–model. In the limit of infinitely heavy vector and axial-vector meson masses integration over the corresponding fields yield (almost) the original version of this model found by Skyrme on purely phenomenological grounds. In Sect. 3.7 we discuss the chiral anomaly using Fujikawa's approach demonstrating the non–invariance of the path integral measure. We will also make a few remarks concerning the chiral anomaly in effective models. Integrating the anomaly for infinitesimally small chiral rotations we derive the famous Wess–Zumino–Witten action. In the last two chapters we will focus on the description of baryons within QFD. First we will shortly review the basic aspects of the picture of baryons as topological solitons, the skyrmions (a short review of a few relevant topological quantities can be found in Appendix B). Plagued with the uncertainties of the stability of skyrmions within the derivative expansion we will use the effective non–local meson theory of Chap. 3 to describe the baryons as (non–topological) solitons. Therefore we have to evaluate the quark determinant exactly which, in fact, can be done for the hedgehog configuration numerically. We will quote numerical results for this soliton including the vector and axialvector mesons. The projection on baryon quantum numbers and fluctuations off the soliton will be discussed in some detail. Strangeness may be included in two ways: either by semiclassical quantization within flavor SU(3) or by considering strange baryons as bound state of kaons and SU(2) solitons. Another possibility to describe baryons within QFD is as a relativistic three quark bound state, see Chap. 5. After giving a functional integral hadronization we will discuss an important building block within this picture: the diquarks. Then it will be demonstrated that the corresponding Fadeev equation becomes

---

* This section may be skipped in a first reading.

a Dirac type equation in a certain approximation. This supplies one with a convenient tool to investigate specific baryon observables. Finally, we will give some conclusions in the final chapter. We will also comment on the question whether the pictures of Chaps. 4 and 5 may be put together in order to arrive at a more complete model of baryons. Especially, it is argued that such a type of hybrid model will open up the possibility to describe the two complementary features of baryons, the collective meson cloud and the bound interacting valence quarks, within one model.

Chapter 2

Reduction of Low–Energy QCD to QFD

2.1 Effective Low–Energy Quark Interaction

In this section we will extract from QCD an effective quark flavor interaction. QCD is defined by the following Lagrangian [IZ85]

$$\mathcal{L}_{\text{QCD}} = \bar{q}(i\gamma^\mu \partial_\mu - m^0)q - \frac{1}{4}(F^a_{\mu\nu})^2 + g\bar{q}\gamma^\mu A_\mu q \qquad (2.1)$$

where q is the quark field which lives in the fundamental representation of the color and flavor group, and $A_\mu = A^a_\mu \lambda^a/2$ is the gluon field with λ^a being the generators of the gauge group. They satisfy the commutation relation $[\lambda^a/2, \lambda^b/2] = if^{abc}\lambda^c/2$. g is the gauge coupling constant. For the quarks color and flavor indices are suppressed. The first term in (2.1) is the free quark Lagrangian with m^0 being the current quark mass. The second term in (2.1) is the Yang–Mills Lagrangian where the field strength $F^a_{\mu\nu}$ is given by

$$F^a_{\mu\nu} = \partial_\mu A^a_\nu - \partial_\nu A^a_\mu + g f^{abc} A^b_\mu A^c_\nu. \qquad (2.2)$$

The term quadratic in the gauge field A^a_μ is due to the non–Abelian nature of the group. For an Abelian group the structure constant vanishes ($f^{abc} = 0$) and therefore this term is absent. The last term in the Lagrangian (2.1) represents the coupling of the gluons to the color current j^a_μ of the quarks

$$\bar{q}\gamma^\mu A_\mu q = A^a_\mu j^\mu_a, \quad j^\mu_a = \bar{q}\frac{\lambda^a}{2}\gamma^\mu q . \qquad (2.3)$$

For the description of strongly interacting quantum systems like hadrons it is not sufficient to study the classical chromodynamics defined by $\mathcal{L}_{QCD}$ (2.1). The full quantum theory (QCD) is defined by the following functional integral (vacuum–to–vacuum transition amplitude)

$$Z_{QCD} = \int \mathcal{D}q \mathcal{D}\bar{q} \int \mathcal{D}A^a_\mu \exp\left(\int d^4x \mathcal{L}_{QCD}\right) . \qquad (2.4)$$

In the functional integral the quark fields q and $\bar{q}$ are described by anticommuting Grassmann variables: $\{q, q\} = \{\bar{q}, \bar{q}\} = \{q, \bar{q}\} = 0$. The integration over the gauge field has to be carried out only over gauge inequivalent orbits. This is usually ensured by using gauge fixing and the Faddeev–Popov method [FP67]. We will assume in the following that this procedure is carried out, however, we will not specify it.**

The physical hadrons are colorless. Their properties seem to be entirely determined by their flavor content. We will therefore try to eliminate the colored gluons and reduce QCD to an effective theory of the flavored quarks. For this purpose we rewrite the generating functional of QCD (2.4) as

$$Z_{QCD} = \int \mathcal{D}q\mathcal{D}\bar{q} \exp\left(\int d^4x \bar{q}(i\gamma^\mu \partial_\mu - m^0)q + \Gamma[j] \right) \qquad (2.5)$$

where

$$\Gamma[j] = \log \int \mathcal{D}A \exp\left(-\frac{1}{4}\int F^2 + g \int A_\mu^a j_a^\mu \right) \qquad (2.6)$$

is an effective action of the color current of the quarks. The gluonic functional integral cannot be evaluated since one can calculate only Gaussian integrals, *i.e.* such integrals which contain the integration variable in the exponent at most the second power. Much work has been concentrated in the past on appropriate approximate integrations over the gluon field, see *e.g.* [CR85, SRAL90].

One possibility to proceed is to expand the effective action in powers of the quark current j_μ^a [CR85, ERV94]

$$\Gamma[j] = \Gamma[j = 0] + g \int \Gamma^{(1)a}{}_\mu j_a^\mu dx_1 + \frac{g^2}{2} \int \Gamma^{(2)}(x_1, x_2)_{\mu_1\mu_2}^{a_1,a_2} \, j_a^{\mu_1} j_b^{\mu_2} dx_1 dx_2 + \dots$$

$$+ \frac{g^n}{n!} \int \Gamma^{(n)}(x_1, x_2 \dots x_n)_{\mu_1 \dots \mu_n}^{a_1 \dots a_n} \, j_{a_1}^{\mu_1}(x_1) \dots j_{a_n}^{\mu_n}(x_n) dx_1 \dots dx_n + \dots .(2.7)$$

The expansion coefficients

$$\Gamma^{(n)}(x_1, x_2 \dots , x_n)_{\mu_1 \dots \mu_n}^{a_1 \dots a_n} = \left(\frac{\delta^{(n)} \Gamma[j]}{\delta j_{\mu_1}^{a_1}(x_1) \dots \delta j_{\mu_n}^{a_n}(x_n)} \right)_{j=0} \qquad (2.8)$$

are given by the one–particle irreducible gluon correlation functions in the pure Yang–Mills theory (*i.e.* in the absence of quarks)

$$\Gamma^{(1)}(x_1)_{\mu_1}^{a_1} = \left\langle A_{\mu_1}^{a_1}(x_1) \right\rangle$$

$$\Gamma^{(2)}(x_1, x_2)_{\mu_1\mu_2}^{a_1,a_2} = \left\langle A_{\mu_1}^{a_1}(x_1) A_{\mu_2}^{a_2}(x_2) \right\rangle \quad - \left\langle A_{\mu_1}^{a_1}(x_1) \right\rangle\left\langle A_{\mu_2}^{a_2}(x_2) \right\rangle \qquad (2.9)$$

** There are recent claims that the correct treatment of Gribov ambiguities [Gr78] has effects on low–energy dynamics [Zw92, Ba92]. We will assume that these effects, if present, can be taken into account by using effective propagators, see below.

and so on where the brackets $\langle \cdots \rangle$ denote the functional average over the gluon field:

$$\langle \ldots \rangle = \frac{\int \mathcal{D}A \ldots \exp\left(-\frac{1}{4}\int F^2\right)}{\int \mathcal{D}A \exp\left(-\frac{1}{4}\int F^2\right)} . \qquad (2.10)$$

Note that none of the gluon correlation functions is gauge or Lorentz invariant. Whereas each term in the expansion is separately invariant under Lorentz and global color transformations only the whole sum (2.7) is invariant under local gauge transformations.

Let us discuss the first few terms of the above current expansion. The zero order term $\Gamma[j=0]$ does not depend on quark fields and is hence an irrelevant constant. The first order term $\Gamma^{(1)}$ gives the expectation value of the gluon field $\langle A_\mu^a \rangle$ which in the absence of external fields vanishes (at least in usual gauges). The leading non-trivial term $\Gamma^{(2)}(x_1, x_2)_{\mu_1 \mu_2}^{a_1 a_2}$ is the gluon correlation function. There are investigations of flavor dynamics and meson physics based on an approximative procedure which consists of neglecting all terms $n \geq 3$ and reasonable assumptions for $\Gamma^{(2)}$ [CR85]. Usually one models this quantity in a way to reproduce confinement and asymptotic freedom of quarks. Recent investigations, however, indicate that the inclusion of non–trivial quark–gluon–vertices is necessary [BP89, SAA91, RW94].

It is interesting to note that the field strength approach (FSA) to non–abelian Yang–Mills theories yields a current–current interaction like the second term in the effective action (2.7) [SRAL90, Re91, AR92]. The underlying idea of the FSA is to integrate out the gluon field and reformulate the theory with the field strength tensor as dynamical field. Putting aside problems connected to gauge fixing this can be done as follows. First, one linearizes the term F^2 in the QCD Lagrangian (2.1) with the help of new auxiliary fields G and χ via the identity

$$1 = \int \mathcal{D}G\, \delta\left(G - \frac{1}{g}F\right) = \int \mathcal{D}G \mathcal{D}\chi \, \exp\left(\frac{i}{2}\int \chi\left(G - \frac{1}{g}F\right)\right) . \qquad (2.11)$$

The tensor fields $G_{\mu\nu}^a$ and $\chi_{\mu\nu}^a$ have the same color and Lorentz structure as the gluon field strength $F_{\mu\nu}^a$. The gluon field enters then the exponential of the generating functional Z_{QCD} (2.4) at most quadratically through the term χF and can hence be integrated out exactly. This yields

$$\Gamma[j] = \log \int \mathcal{D}G \mathcal{D}\chi \exp\left(-\frac{1}{4}\int d^4x\, G_{\mu\nu}^a G_{\mu\nu}^a + \frac{i}{2}\int d^4x\, \chi_{\mu\nu}^a G_{\mu\nu}^a + \tilde{S}(\chi, j)\right),$$

$$\tilde{S}(\chi, j) = -\frac{1}{2}\mathrm{Tr}\log(ig\hat{\chi}) - \frac{i}{2g}\int d^4x\, \chi_{\mu\nu}^a V_{\mu\nu}^a + \int d^4x\, j_\mu^a V_\mu^a . \qquad (2.12)$$

Here

$$\hat{\chi}_{\mu\nu}^{ab} = f^{abc}\chi_{\mu\nu}^c = i(T_c)^{ab}\chi_{\mu\nu}^c \qquad (2.13)$$

is the group valued tensor field in the regular representation of the gauge group generators $T_c, c = 1, 2, ..., N_c^2 - 1$. Furthermore, $V_{\mu\nu}^a$ is the field strength tensor of the 'dynamical' vector field

$$V_\mu^a = J_\mu^a - i(\hat{\chi}^{-1})_{\mu\nu}^{ab} \, g \, j_\nu^b \, , \quad J_\mu^a = (\hat{\chi}^{-1})_{\mu\nu}^{ab} \, \partial_\lambda \chi_{\lambda\nu}^b \tag{2.14}$$

which transforms under gauge transformations precisely as the original gluon field A_μ^a as the reader may easily check from its definition. Consequently the corresponding field strength $V_{\mu\nu}^a$ transforms like $F_{\mu\nu}^a$. The integral over the field G is Gaussian. Performing this integral one obtains an effective tensor field action

$$\Gamma[j] = \log \int \mathcal{D}\chi e^{S(\chi,j)} \, ,$$

$$S(\chi, j) = -\frac{1}{4} \int d^4x \chi_{\mu\nu}^a \chi_{\mu\nu}^a + \tilde{S}(\chi, j) \, . \tag{2.15}$$

which is equivalent to the effective action (2.6). Note that the quantity $\mathrm{Tr} \log(ig\hat{\chi})$ which stems from the integration over gluon field needs regularization. This introduces an energy scale μ which breaks the scale invariance of the classical Lagrangian (2.1) in an anomalous fashion.

One may verify that the action (2.15) is stationary for non–vanishing, albeit purely imaginary, space–time independent field configurations:

$$\bar{\chi} = -iG. \tag{2.16}$$

Therefor this effective action describes a non–perturbative vacuum state already at the tree level of the χ field.

If one is interested in the effective quark interaction it is instructive to rewrite the generating functional (2.5) using eq. (2.15)

$$Z_{QCD} = \int \mathcal{D}\chi e^{S(\chi,j=0)} Z_f[\chi] \, ,$$

$$Z_f[\chi] = \int \mathcal{D}q\mathcal{D}\bar{q} \exp\left(\int d^4x(\bar{q}(i\slashed{\partial} - m_0)q + j_\mu^a J_\mu^a + \frac{g}{2}j_\mu^a((i\hat{\chi})^{-1})_{\mu\nu}^{ab} j_\nu^b) \right) \, . \tag{2.17}$$

For a given χ the fermionic transition amplitude $Z_f[\chi]$ describes a quark system in an 'external' color vector field J_μ^a and interacting via a local but dynamical two–body force $((i\hat{\chi})^{-1})_{\mu\nu}^{ab}(x)$. One may expand χ around the stationary phase points $\bar{\chi}$ of $S(\chi, j = 0)$

$$\chi = \bar{\chi} + \varphi \, . \tag{2.18}$$

The generating functional is then given by

$$Z_{QCD} = \sum_{\bar{\chi}} \int \mathcal{D}\varphi \exp\left(S(\bar{\chi}) - \frac{1}{2}\int d^4x\,\varphi D_\varphi^{-1}\varphi\right) Z_f[\bar{\chi},\varphi]\,,$$

$$Z_f[\bar{\chi},\varphi] = \int \mathcal{D}q\mathcal{D}\bar{q}\exp\Bigg(\int d^4x(\bar{q}(i\slashed{\partial} - m_0)q)$$

$$+\frac{g^2}{2}\int d^4x\,d^4y j_\mu^a(x)(\Gamma^{(2)}(x,y))_{\mu\nu}^{ab}j_\nu^b(y)\Bigg) \tag{2.19}$$

where D_φ is the propagator of the φ-field and $\Gamma^{(2)}(x,y)$ is now an effective non-local four–point quark interaction in the background field $\bar{\chi} = -iG$. For small momenta we obtain a local contact interaction

$$\Gamma^{(2)}(x,y)_{\mu\nu}^{ab} = \delta(x-y)(\hat{G}^{-1})_{\mu\nu}^{ab} \tag{2.20}$$

similar to the one known from Nambu–Jona-Lasinio type of models [NJL61]. However, its strength and Lorentz and color structure is determined by the classical gluonic vacua. In the opposite limit of large momenta $\Gamma^{(2)}$ behaves like the perturbative gluon propagator [Re90a, Sm94]. So, at large momenta the effective quark interaction becomes identical to the perturbative one gluon exchange.

Using the low–energy expression (2.20) is equivalent to neglect the influence of the gluonic fluctuations φ on the effective quark interaction. This interaction is then described by $(\hat{G}^{-1})_{\mu\nu}^{ab}$ which is a color and Lorentz matrix of the form

$$G_{\mu\nu}^{ab} = \kappa\left(b^{ab}g_{\mu\nu} + c_k^{ab}\eta_{\mu\nu}^k\right)\,. \tag{2.21}$$

Here $g_{\mu\nu}$ is the metric tensor, the $\eta_{\mu\nu}^k$ are the 't Hooft symbols [tH76], i.e. four–dimensional $s = \frac{1}{2}$ representations of SU(2), and κ is determinated by the strength of the gluon condensation, $< (F_{\mu\nu})^2 > \propto \kappa^2$. The matrices b and c_k are symmetric and antisymmetric color matrices, respectively. For the gauge group SU(2) $b^{ab} = \delta^{ab}$ whereas for SU(3) there are slight deviations of b from the unit matrix [AR92]. In the following we will assume that the deviations of this matrix from the unit matrix are of little importance for the evaluation of physical quantities which are color singlet, i.e. we will simply assume***

$$\Gamma^{(2)}(x,y)_{\mu\nu}^{ab} = \kappa\delta(x-y)\delta^{ab}g_{\mu\nu}. \tag{2.22}$$

With these simplifications the QCD generating functional (2.5) is approximated by

$$Z_{QCD} \approx \int \mathcal{D}q\mathcal{D}\bar{q}\exp\left(\int d^4x \mathcal{L}_{QFD}\right)\,, \tag{2.23}$$

*** In fact, this result is obtained when one averages over all gauge and Lorentz equivalent "background" fields $\bar{\chi} = -iG$, see [Re91, AR92].

where the QFD Lagrangian is given by

$$\mathcal{L}_{QFD} = \bar{q}(i\gamma^\mu \partial_\mu - m^0)q + \mathcal{L}_{int} ,$$

$$\mathcal{L}_{int} = \frac{-\kappa g^2}{2} j_\mu^a j^{a\mu}. \tag{2.24}$$

It describes a system of quarks interacting via a two–body–force. The "effective" low–energy Lagrangian is very much remeniscent to an extended NJL model [NJL61], as we will see shortly.

2.2 Invariance Properties of QCD and QFD

Before proceeding further with the effective quark theory (2.24) let us study its symmetries and compare them with the symmetries of the QCD Lagrangian (2.1). Both $\mathcal{L}_{QCD}$ and $\mathcal{L}_{QFD}$ are obviously invariant under Lorentz transformations. Furthermore, by construction, $\mathcal{L}_{QCD}$ is gauge invariant:

$$q \to U_c q , \quad \bar{q} \to \bar{q} U_c^\dagger$$

$$A_\mu \to U_c A_\mu U_c^\dagger - \frac{1}{g} U_c i \partial_\mu U_c^\dagger ,$$

$$U_c(x) = \exp\big(i\theta^a(x)(\frac{\lambda^a}{2})_c\big). \tag{2.25}$$

This invariance is also present for the exact effective quark theory defined by (2.5) and (2.6) but is lost due to the low–energy approximation (2.20). Global color invariance, however, is fully respected by the effective quark theory (2.24).

The QCD Lagrangian contains no dimensionful parameter and is hence scale invariant.[†] On the other hand, the effective quark theory (2.24) with the low–energy limit (2.22) for $\Gamma^{(2)}$ contains the parameter κ of dimension (energy)^2. This dimensionful constant reflects the anomalous breaking of scale invariance when one integrates (averages) over the gluon field. Furthermore, $\mathcal{L}_{QCD}$ and $\mathcal{L}_{QFD}$ are invariant under global flavor rotations

$$q \to U_V q, \quad \bar{q} \to \bar{q} U_V^\dagger, \quad U_V = \exp(i\theta_V), \quad \theta_V = \theta_V^a \left(\frac{\lambda^a}{2}\right)_F \tag{2.26}$$

where $\left(\lambda^a/2\right)_F$ denotes the generators of the flavor group $U(N_f)$, N_f being the number of flavors. For reasons which will become clear later this flavor rotation is referred to as vector flavor transformation and we will denote the flavor group as $U_V(N_f)$. Note that the gluons are flavor blind and are hence not effected by flavor transformations.

Besides the above discussed exact symmetries the QCD Lagrangian has an approximate symmetry which becomes exact for $m^0 \to 0$: **chiral symmetry**.

[†] In a strict sense this is only true if the current masses of the quarks are neglected. However, the limit of vanishing current masses is smooth in QCD.

A chiral transformation on the quarks is like the above discussed flavor rotation except that there is an additional γ_5 multiplying the flavor generators

$$q \to U_A q, \quad U_A = \exp\left(i\gamma_5\theta_A\right), \quad \theta_A = \theta_A^a \left(\frac{\lambda^a}{2}\right)_F . \tag{2.27}$$

Due to the presence of the γ_5 this transformation is also called axial–vector transformation since local axial transformations induce an axial–vector field. Exploiting $\{\gamma_5, \gamma^0\} = 0$ the conjugate Dirac field $\bar{q} = q^\dagger\gamma^0$ transforms under the axial transformations (2.27) as

$$\bar{q} \to \bar{q}U_A . \tag{2.28}$$

Note that contrary to vector transformation (2.26) the axial–vector transformation (2.27) and (2.28) on the fields q and $\bar{q}$ is represented by the same matrix and not by the hermitean conjugate. The vector and the axial–vector transformation are usually combined, which is then referred to as chiral transformation and the corresponding symmetry

$$U_V(N_f) \otimes U_A(N_f) \tag{2.29}$$

is called chiral symmetry. Using $\gamma_5^2 = 1$ one can define two projection operators

$$P_R = \frac{1}{2}(1 + \gamma_5), \quad P_L = \frac{1}{2}(1 - \gamma_5) \tag{2.30}$$

with the following properties

$$\begin{aligned} P_R + P_L &= 1, \\ P_R^2 &= P_R, \quad P_L^2 = P_L, \\ P_R P_L &= P_L P_R = 0. \end{aligned} \tag{2.31}$$

Since $\gamma_5 = P_R - P_L$ and $U_A = (U_V)^{\gamma_5}$ we can rewrite the chiral transformation matrices as

$$\begin{aligned} U_V(\theta_V)U_A(\theta_A) &= U_V(\theta_V)\left(U_V(\theta_A)\right)^{\gamma_5} \\ &= U_V(\theta_V)^{P_R+P_L}U_V(\theta_A)^{P_R-P_L} . \end{aligned} \tag{2.32}$$

Using the fact that P_L and P_R are orthogonal projectors (see 2.31) and hence commute we obtain immediately

$$\begin{aligned} U_V(\theta_V)U_A(\theta_A) &= (P_R e^{i\theta_V} + P_L e^{i\theta_V})(P_R e^{i\theta_A} + P_L e^{-i\theta_A}) \\ &= P_R e^{i\theta_R} + P_L e^{i\theta_L} \\ &= P_R U_V(\theta_R) + P_L U_V(\theta_L) . \end{aligned} \tag{2.33}$$

In the special case of an abelian group $U(N_f = 1)$ one has

$$\theta_R = \theta_V + \theta_A , \quad \theta_L = \theta_V - \theta_A , \tag{2.34}$$

whereas for a non–abelian group the relation between $\theta_{R,L}$ and $\theta_{V,A}$ is more complicated. Here $\theta_{R,L}$ is defined by $U_V(\theta_R) = U_V(\theta_V)U_V(\theta_A)$ and $U_V(\theta_L) = U_V(\theta_V)U_V(-\theta_A)$.

The left– and right–handed flavor transformations act independently on the right– and left–handed quark fields defined by

$$q_R = P_R q, \quad q_L = P_L q. \tag{2.35}$$

In fact under a chiral $U_V(N_f) \otimes U_A(N_f)$ transformation one obtains using eqs. (2.26) and (2.27)

$$q_L \to U_V(\theta_L)q_L,$$
$$q_R \to U_V(\theta_R)q_R. \tag{2.36}$$

Since $\{\gamma_\mu, \gamma_5\} = 0$ the conjugate fields

$$\bar{q}_L = \bar{q}P_R , \quad \bar{q}_R = \bar{q}P_L \tag{2.37}$$

transform as

$$\bar{q}_L \to \bar{q}_L U_V^\dagger(\theta_L) ,$$
$$\bar{q}_R \to \bar{q}_R U_V^\dagger(\theta_R). \tag{2.38}$$

For $m^0 = 0$ the QCD Lagrangian (2.1) is invariant under this global chiral $U_V(N_f) \otimes U_A(N_f)$ transformation. Let us explicitly demonstrate this: Since this transformation is space–time independent and color blind it commutes with the momentum operator ∂_μ and the color group generator λ^a. It suffices therefore to show the invariance of the current $\bar{q}\gamma^\mu q$ under the chiral transformation (2.33)

$$\begin{aligned}
\bar{q}\gamma^\mu q &= \bar{q}(P_L + P_R)\gamma^\mu(P_L + P_R)q \\
&= \bar{q}P_L\gamma^\mu P_R q + \bar{q}P_R\gamma^\mu P_L q \\
&= \bar{q}_R\gamma^\mu q_R + \bar{q}_L\gamma^\mu q_L .
\end{aligned} \tag{2.39}$$

Only the diagonal terms survive since $\gamma^\mu P_R = P_L\gamma^\mu$. From (2.36) and (2.38) it is trivial to see that the diagonal terms remain invariant. The current quark mass term in $\mathcal{L}_{QCD}$ spoils the chiral symmetry even if m^0 is flavor diagonal

$$\mathcal{L}_{mass} = -\bar{q}m^0 q = -\bar{q}(P_L + P_R)m^0(P_L + P_R)q = -(\bar{q}_R m^0 q_L + \bar{q}_L m^0 q_R). \tag{2.40}$$

The physical meaning of chiral symmetry becomes obvious if one considers free massless fermions (say quarks) which have this symmetry as we have seen above. Such fields $q(x)$ satisfy the Dirac equation in the momentum representation

$$\not{p}\, q(x) = 0 \quad \text{where} \quad \not{p} = \gamma_\mu p^\mu = \gamma^0 E - \boldsymbol{\gamma p} . \tag{2.41}$$

Since

$$\boldsymbol{\gamma} = \gamma^0 \boldsymbol{\alpha} , \quad \boldsymbol{\alpha} = \gamma_5 \otimes \boldsymbol{\sigma} \tag{2.42}$$

the free Dirac equation can be rewritten as

$$(E - \boldsymbol{\sigma} \boldsymbol{p} \gamma_5)\, q(x) = 0. \tag{2.43}$$

Furthermore, for a massless particle $E = |\boldsymbol{p}|$. Hence, separating right– and left–handed fields in the Dirac equation yields

$$\frac{\boldsymbol{\sigma} \boldsymbol{p}}{|\boldsymbol{p}|}\, q_R(x) = q_R(x)$$

$$\frac{\boldsymbol{\sigma} \boldsymbol{p}}{|\boldsymbol{p}|}\, q_L(x) = -q_L(x)\,. \tag{2.44}$$

The operator $\frac{\boldsymbol{\sigma} \boldsymbol{p}}{|\boldsymbol{p}|}$ acting on the fields $q_{L,R}$ is the helicity operator. This shows that right– and left–handed massless fermions are eigenstates of the helicity (or chirality) with eigenvalues ± 1.

The chiral symmetry is one of the most important symmetries in low–energy hadron physics. It has far reaching consequences and is manifest in the low–lying meson spectrum as we will see later. Furthermore, it was the guiding principle in the construction of the very successful phenomenological meson Lagrangians of the sixties.

On the strong interaction scale the up and down current quark masses are very small. This is often also assumed for the strange current quark mass. Hence, the limit $m^0 \to 0$ should be a good approximation for the light quarks. As we have seen above in this limit the QCD Lagrangian obeys chiral symmetry. Therefore it should be visible in the strong interaction particle spectrum. Since particles with opposite helicity are related by a parity transformation in a chirally symmetric world the hadrons should come in parity doublets. However, in the hadron spectrum we do not see any such degeneracy. Therefore one assumes that chiral symmetry is not realized in the ground state, *i.e.* chiral symmetry is spontaneously broken with the pseudoscalar mesons (pions, kaons and etas) appearing as the corresponding Goldstone bosons. The concept of spontaneous broken chiral symmetry is the cornerstone in the understanding of the light mesons. In the following we will demonstrate that in the QFD defined by eq. (2.24) chiral symmetry is in fact spontaneously broken. Furthermore, QFD describes the wealth of low–energy meson physics reasonably well in a very efficient way. It is also a good starting point for the descriptions of baryons.

2.3 Fierz–Transformation of the Effective Quark Interaction

The effective quark interaction $\mathcal{L}_{int}$ (2.24) is a local two–body force similar to the current–current interaction of the Fermi model of weak interaction. However, $\mathcal{L}_{int}$ contains the color octet flavor singlet currents of the quarks. As already stated, in the low–energy hadron spectra the flavor quantum numbers are seen whereas all hadrons are color singlets. It is therefore advantageous to rewrite the effective quark interaction in a way that it acts in physical channels. This can be straightforwardly done by a Fierz transformation [CRP87, Re90]. Naïvely, one would transform the effective quark interaction $\mathcal{L}_{int}$ only in the fermion–antifermion channel thereby creating color singlet and octet states with meson quantum numbers as can be seen from the relation

$$\left(\frac{\lambda^a}{2}\right)_{ij}\left(\frac{\lambda^a}{2}\right)_{kl} = \frac{1}{2}\left(1 - \frac{1}{N_c^2}\right)\delta_{il}\delta_{kj} - \frac{1}{N_c}\left(\frac{\lambda^a}{2}\right)_{il}\left(\frac{\lambda^a}{2}\right)_{kj} . \tag{2.45}$$

However, in nature no color octet mesons exist. Looking at the relative signs in (2.45) the reason becomes immediately clear: The interaction in color singlet and octet channel differ in sign, *i.e.* if it is attractive in the singlet channel (as in nature) it is repulsive in the octet channel and no bound state exists in this channel. In order to resolve this puzzle we are going to have a closer look at the baryons, *i.e.* the three–quark color singlet states. Coupling two quarks in the fundamental triplet representation 3_c gives either a sextet 6_c or an antitriplet $\bar{3}_c$. As only the antitriplet diquark state can be coupled with the third quark being in 3_c to a color singlet we can conclude immediately that in a baryon every pair of quarks is in a color antitriplet state. Note also that the diquark in $\bar{3}_c$ is antisymmetric and in 6_c symmetric in the color quantum numbers of the underlying quarks. Furthermore, choosing the sign of κ such that $\mathcal{L}_{int}$ is attractive in the color singlet state our effective quark interaction is attractive in the $\bar{3}_c$ diquark state and repulsive in the 6_c diquark channel. As we want to describe bound states we would like to rewrite $\mathcal{L}_{int}$ in such a way that it contains only attractive interactions. Using the totally antisymmetric tensor ϵ_{ijk} as a representation of $\bar{3}_c$ we realize that this can be done with the help of the following relation:

$$\left(\frac{\lambda^a}{2}\right)_{ij}\left(\frac{\lambda^a}{2}\right)_{kl} = \frac{1}{2}\left(1 - \frac{1}{N_c}\right)\delta_{il}\delta_{kj} + \frac{1}{2N_c}\epsilon_{mik}\epsilon_{mlj} . \tag{2.46}$$

Guided by this physical principle of choosing the attractive channels of our effective interaction the way of rewriting is **uniquely** determined. Of course, we have also to Fierz the flavor and Dirac quantum numbers in both channels present in (2.46). Let us first start with the flavor indices. In contrast to the color group $SU(N_c)$ the flavor group is $U(N_f)$, *i.e.* the flavor generators $\lambda^a/2$ include $\lambda^0/2$ defined as $\mathbb{1}/\sqrt{2N_f}$. Therefore, one can write the completeness relation in the meson channel

$$\delta_{ij}\delta_{kl} = 2\left(\frac{\lambda^0}{2}\right)_{il}\left(\frac{\lambda^0}{2}\right)_{kj} + 2\sum_{a=1}^{N_f^2-1}\left(\frac{\lambda^a}{2}\right)_{il}\left(\frac{\lambda^a}{2}\right)_{kj} = 2\sum_{a=0}^{N_f^2-1}\left(\frac{\lambda^a}{2}\right)_{il}\left(\frac{\lambda^a}{2}\right)_{kj} \tag{2.47}$$

In order to be specific and simplify notation let us use $N_f = 3$ [‡]. The decomposition (2.47) then tells us that mesons occur as nonets (singlet + octet) under

---

[‡] Most of the things said in the following can straightforwardly be generalized to arbitrary N_f.

flavor. For the vector mesons ρ, ω, ϕ and K^* or the pseudoscalar mesons π, K, η and η' these nonets are quite easily identified.[§]

The rearranging of flavor indices in the diquark channel

$$\delta_{ij}\delta_{kl} = \frac{1}{2}(\delta_{ij}\delta_{kl} - \delta_{il}\delta_{kj}) + \frac{1}{2}(\delta_{ij}\delta_{kl} + \delta_{il}\delta_{kj})$$

$$= 2\sum_{m=1}^{3} A_{ik}^m A_{lj}^m + 2\sum_{n=1}^{6} S_{ik}^n S_{lj}^n \qquad (2.48)$$

tells us that diquarks are either in the antisymmetric $\bar{3}_F$ or the symmetric 6_F representation. Note that $A_{ik}^m = i\epsilon_{mik}$ is a possible representation of $\bar{3}_F$ to parallel the expression in (2.46) where we did choose the completely antisymmetric tensor ϵ from the beginning as representation of $\bar{3}_c$. However, for the flavor quantum numbers it is more convenient to choose the matrices A as $A^1 = \lambda^7/2, A^2 = -\lambda^5/2$ and $A^3 = \lambda^2/2$ which differ from the ϵ_{mik} besides the factor 2 also by a sign. The matrices S may then be chosen as $S^1 = \lambda^0/2$, $S^2 = \lambda^3/2$, $S^3 = \lambda^8/2$, $S^4 = \lambda^1/2$, $S^5 = \lambda^4/2$ and $S^6 = \lambda^6/2$.

Finally, we have to rearrange Dirac indices. For the meson channel this transformation is given in most textbooks on field theory, see *e.g.* [IZ85],

$$(\gamma_\mu)_{ij}(\gamma^\mu)_{kl} = \delta_{il}\delta_{kj} + (i\gamma_5)_{il}(i\gamma_5)_{kj}$$

$$- \frac{1}{2}\left((\gamma_\mu)_{il}(\gamma^\mu)_{kj} + (\gamma_\mu\gamma_5)_{il}(\gamma^\mu\gamma_5)_{kj}\right) . \qquad (2.49)$$

In the diquark channel things are more complicated. Let us first recall a few definitions. Given the spinor q the charge conjugated spinor is given by

$$q^c = C\bar{q}^T = C(\gamma^0)^T q^* \qquad (2.50)$$

where C is the charge conjugation matrix. It satisfies the following relations [¶]

$$C^\dagger = C^T = C^{-1} , \quad C^2 = -1 \qquad (2.51)$$

and

$$C\gamma^\mu C = (\gamma^\mu)^T. \qquad (2.52)$$

These properties are used for reordering the Dirac spinors in the effective interaction $\mathcal{L}_{int}$ (2.24)

$$(\gamma_\mu)_{ij}(\gamma^\mu)_{kl} = (\gamma_\mu)_{ij}\left((\gamma^\mu)^T\right)_{lk}$$

$$= (\gamma_\mu)_{ij}(C\gamma^\mu C)_{lk}$$

$$= (\gamma_\mu)_{ij}C_{lm}(\gamma^\mu)_{mn}C_{nk} . \qquad (2.53)$$

[§] There are mixings between the neutral mesons in these nonets which in the case of the vector mesons can be understood in terms of the current quark masses only whereas in the pseudoscalar case also some additional interactions related to the chiral anomaly are needed in order to explain the η'–mass and the $\eta - \eta'$–mixing.
[¶] In the Dirac representation $C = i\gamma^2\gamma^0$.

Applying now (2.49) in (2.53) and writing contracted indices again as matrix multiplication one obtains

$$(\gamma_\mu)_{ij}(\gamma^\mu)_{kl} = C_{ik}C_{lj} + (i\gamma_5 C)_{ik}(C i\gamma_5)_{lj}$$
$$- \frac{1}{2}\left((\gamma_\mu C)_{ik}(C\gamma^\mu)_{lj} + (\gamma_\mu \gamma_5 C)_{ik}(C\gamma^\mu \gamma_5)_{lj}\right). \qquad (2.54)$$

Now we are prepared to rearrange the interaction $\mathcal{L}_{int}$ ($N_c = N_f = 3$ is used)

$$\mathcal{L}_{int} = -\kappa \frac{g^2}{2} j_\mu^a j^{\mu a}$$
$$= \frac{g^2}{3}\kappa\, (\bar{q}\Lambda_\alpha q \bar{q}\Lambda^\alpha q + \bar{q}\Sigma_\alpha q^c \bar{q}^c \Sigma^\alpha q)$$
$$=: \mathcal{L}_{int}^{q\bar{q}} + \mathcal{L}_{int}^{qq} \qquad (2.55)$$

where $j_\mu^a = \bar{q}\gamma_\mu(\lambda^a/2)_C 1\!\!1_F q$ is the color octet flavor singlet quark current of sect. 2.1 and the matrices Λ_α and Σ_α are tensor products of color, flavor and Dirac matrices

$$\Lambda_\alpha = 1\!\!1_C \otimes (\frac{\lambda^A}{2})_F \otimes \Gamma_\alpha\,, \quad A = 0,\ldots 8\,, \qquad (2.56)$$

and

$$\Sigma_\alpha = (\frac{i}{\sqrt{2}}\epsilon^a)_C \otimes t_F^A \otimes \Gamma_\alpha\,, \quad a = 1,2,3; \quad t_F^A \epsilon \{A^m, S^n\}. \qquad (2.57)$$

In both cases the Dirac matrices are given by

$$\Gamma_\alpha \in \{1\!\!1, i\gamma_5, \frac{i}{\sqrt{2}}\gamma_\mu, \frac{i}{\sqrt{2}}\gamma_\mu\gamma_5\}\,. \qquad (2.58)$$

The matrices Λ_α and Σ_α may be interpreted as the meson and diquark vertices, respectively, of QFD.

The first term in the Lagrangian (2.55) gives the interaction in the color singlet (1_c) quark-antiquark channel while the second one describes diquarks in the $\bar{3}_c$ representation. Note that $\mathcal{L}_{int}^{q\bar{q}}$ and $\mathcal{L}_{int}^{qq}$ are separately chirally invariant. This will allow us to discuss the two different terms independently from each other.

A crucial difference between the quark–antiquark and the diquark vertices is due to the Pauli principle which acts restrictively in the diquark channel but is inactive in the quark–antiquark channel. While in the latter the Dirac matrix can occur with any flavor vertex in the former the Pauli principle allows only for certain combinations of Dirac and flavor vertices. We will return to this issue in Chap. 5 where diquarks are discussed in detail.

Finally, we want to emphasize that for general $N_c \neq 3$ the factors $1/3$ in the interaction terms $\mathcal{L}_{int}^{q\bar{q}}$ and $\mathcal{L}_{int}^{qq}$ are given by $(N_c - 1)/2N_c$ and $1/N_c$, respectively. Thus in the large-N_c limit the diquark term $\mathcal{L}_{int}^{qq}$ is suppressed with respect to $\mathcal{L}_{int}^{q\bar{q}}$ by a factor $1/N_c$. In this limit there are no bound diquark (($N_c - 1$)-quark) states and consequently baryons as well as the mesons must be exclusively generated from the attractive $(q\bar{q})_{1c}$ interactions $\mathcal{L}_{int}^{q\bar{q}}$. Therefore we will concentrate on this term in the next two chapters.

Chapter 3

The Effective Meson Theory

3.1 Functional Integral Bosonization of the Quark–Antiquark Interaction

As stated we will first discuss the quark-antiquark interaction $\mathcal{L}_{int}^{q\bar{q}}$ separately from the quark-quark interaction $\mathcal{L}_{int}^{qq}$. Keeping only $\mathcal{L}_{int}^{q\bar{q}}$ we obtain a model whose generic form predates QCD. It was as early as 1961 that Y. Nambu and G. Jona-Lasinio studied a model with a chirally symmetric local four-point interaction [NJL61]. They interpreted the underlying fermion fields as nucleons. They found dynamical breaking of chiral symmetry and pseudoscalar Goldstone bosons. In recent years there was much renewed interest in models of the Nambu-Jona-Lasinio (NJL) type, see *e.g.* [ER86, RA88, BJM88, Re90, KLVW90]. Recent reviews are [ERV94, HK94]. This revival of the NJL model was triggered by the large N_c considerations of 't Hooft [tH74] and Witten [Wi79] who demonstrated that QCD becomes for $N_c \rightarrow \infty$ an effective theory of weakly interacting mesons and glueballs and baryons appear as solitons of the meson fields. Unfortunately, the effective meson theory of large N_c QCD cannot be constructed explicitly. The primary aim of the investigations using the NJL model was to understand the emergence of the effective meson theory [ER86] and in particular the appearance of baryons as solitons out of the quark flavor dynamics. By functional integral techniques the NJL model was rewritten in terms of physical hadron fields [Re90]. The resulting effective meson theory proved to describe quite successfully meson properties [ER86] like masses, decay constants, scattering lengths and so on. It also explained the emergence of baryons as chiral solitons. Especially, it contains the well–known Skyrme model [Sk61, Ho93] as a certain low–energy limit [RD89], see Sect. 3.6. Recently, there have been efforts to explicitly describe baryons within NJL models, either as chiral solitons of non-linear meson field configurations [RW88, MGG89, Al90, AR91, WAW93] or as bound states of three constituent quarks [Re90, BAR92].

Let us now derive the effective meson theory corresponding to the Lagrangian

$$\mathcal{L}_{NJL} = \mathcal{L}_0 + \mathcal{L}_{int}^{q\bar{q}} \tag{3.1}$$

by the use of functional integral techniques. In order to make some of the following discussion more transparent we write $\mathcal{L}_{int}^{q\bar{q}}$ as

$$\mathcal{L}_{int}^{q\bar{q}} = \frac{1}{2}\bar{q}\Lambda_\alpha q Q^{\alpha\beta}\bar{q}\Lambda_\beta q \tag{3.2}$$

and by comparison with eq. (2.55) one obtains

$$Q^{\alpha\beta} = \frac{2}{3}g^2\kappa\delta^{\alpha\beta} \tag{3.3}$$

for the NJL-type model (3.1).

The aim is to rewrite the quark (fermion) theory (3.1) into an effective meson (boson) theory. Therefore we introduce an auxiliary field $\eta = \eta_\alpha\Lambda_\alpha$ via the identity

$$\exp\left(\frac{1}{2}\int\bar{q}\Lambda_\alpha q Q^{\alpha\beta}\bar{q}\Lambda_\beta q\right) = \frac{1}{\sqrt{\text{Det}2\pi Q}}$$
$$\int\mathcal{D}\eta\exp\left(-\frac{1}{2}\int\eta_\alpha(Q^{-1})^{\alpha\beta}\eta_\beta - \int\eta_\alpha\bar{q}\Lambda_\alpha q\right). \tag{3.4}$$

The symbol "Det" means here the functional determinant as well as the determinant over color, flavor and Dirac indices.

Using eq. (3.4) the generating functional $Z_{NJL} = \int\mathcal{D}q\mathcal{D}\bar{q}\exp(\mathcal{L}_{NJL})$ may be written as

$$Z_{NJL} = (\text{Det}(2\pi Q))^{-1/2}\int\mathcal{D}\eta\exp\left(-\frac{1}{2}\int\eta Q^{-1}\eta\right)Z_F[\eta]$$
$$Z_F[\eta] = \int\mathcal{D}q\mathcal{D}\bar{q}\exp\left(\int\bar{q}(i\partial\!\!\!/ - m_0 - \eta)q\right). \tag{3.5}$$

For the interpretation of $Z_F[\eta]$ we note that it is equivalent to

$$Z_F[\eta] = \lim_{T\to\infty}\langle 0|e^{-\hat{h}T}|0\rangle \tag{3.6}$$

where $\hat{h} = q^+hq$ is an one-particle Dirac Hamiltonian and T may be interpreted as a large Euclidean time interval. In Minkowski space this Hamiltonian is given by

$$h = \boldsymbol{\alpha}\boldsymbol{p} + \beta(m_0 + \eta) \tag{3.7}$$

as can be verified from the relation $i\partial\!\!\!/ - m_0 - \eta = \beta(i\partial_t - h)$. From (3.6) one sees that $Z_F[\eta]$ is the quantum transition amplitude of a non-interacting quark system in an external field $\eta = \eta_\alpha\Lambda_\alpha$. In order to obtain the generating functional

Z_{NJL} we have to integrate over all possible configurations of the η-field with a Gaussian weight

$$Z_{NJL} = \int \mathcal{D}\eta\, P[\eta] Z_F[\eta]$$

$$P[\eta] = (\mathrm{Det}(2\pi Q))^{-1/2} \exp\left(\frac{1}{2} \int d^4x\, \eta Q^{-1}\eta\right). \tag{3.8}$$

The term $(\mathrm{Det} 2\pi Q)^{-1/2}$ is an irrelevant constant and will be omitted in the following[*]. The width of the Gaussian $P[\eta]$ is $\sqrt{Q}$, *i.e.* it is given by the effective quark interaction. For weak coupling, *i.e.* Q small, fluctuations of the meson field η are restricted to small values whereas for strong coupling, *i.e.* Q large, one expects large amplitude fluctuations of η. Especially, it is easy to find the limit of a free quark theory because

$$\lim_{Q\to 0} P[\eta] = \lim_{Q\to 0} \left(\sqrt{\mathrm{Det}(2\pi Q)}\right)^{-1} \exp\left(-\frac{1}{2}\int \eta Q^{-1}\eta\right) = \delta(\eta) \tag{3.9}$$

and therefore $Z_{NJL} = Z_F[\eta = 0]$ in this limit.

Note that the weight $P[\eta]$ might change as the quark (fermion) systems respond to an external field, *e.g.* if the Dirac see becomes polarized due to Debye-screening.

The vacuum expectation value (VEV) of the auxiliary field η is found from the stationary point of the action given in (3.5):

$$\langle \eta_\alpha \rangle = -Q^{\alpha\beta}\langle \bar{q}\Lambda_\beta q \rangle. \tag{3.10}$$

This demonstrates that η is a composite collective field and that its VEV is proportional to the two-body force. This situation is well-known from many-particle theories [Re80], *c.f.* RPA with separable interaction [FW71, RS80].

In eq. (3.5) the quark field appears bilinear in the exponent and can therefore be integrated out. Additionally, performing a shift in the integration variable $\eta \to \eta - m^0$ and using that $\mathrm{Det} = \exp\mathrm{Tr}\log$ one obtains

$$Z_{NJL} = \int \mathcal{D}\eta\, e^{\mathcal{A}[\eta]}$$

$$\mathcal{A}[\eta] = -\frac{1}{2}\int d^4x\, (\eta - m^0)Q^{-1}(\eta - m^0) + \mathrm{Tr}\log(i\!\!\not{\partial} - \eta). \tag{3.11}$$

Note that the action $\mathcal{A}[\eta]$ is a non-linear, even non-polynomial function of the meson field η. Even more, the term $\mathrm{Tr}\log(i\!\!\not{\partial} - \eta)$ is non-local. The action $\mathcal{A}[\eta]$ is, however, exactly equivalent to the underlying NJL model defined by the Lagrangian (3.1). On the other hand, the generating functional (3.11) has the advantage that it may be treated in a semiclassical approximation because the VEV of the bosonic field η can be different from zero, and in general will be, whereas the VEV of a fermionic quark field necessarily vanishes. As can be seen from eq. (3.10) $\langle\eta\rangle \neq 0$ implies a non-vanishing quark condensate $\langle\bar{q}q\rangle$.

---

[*] In more elaborate theories where one also takes into account the gluons at least in a semiclassical fashion this term might be important.

3.2 Small–Amplitude Expansion of the Action

Variation of the effective action $\mathcal{A}[\eta]$ (3.11) yields the Dyson–Schwinger equation

$$\eta_\alpha(x) = -Q_{\alpha\beta}\mathrm{tr}(G_\eta(x,x)\Lambda_\beta). \tag{3.12}$$

The solution of this equation determines the VEV of the meson field η. G_η is hereby the quark propagator in the background of the η-field. It is defined by

$$G_\eta^{-1}(x,y) = (i\partial\!\!\!/ - \eta)\delta(x-y). \tag{3.13}$$

For later use we introduce also the matrix notation

$$\eta_{ij} = \eta_\alpha \Lambda_{ij}^\alpha$$
$$Q_{ij,kl} = \Lambda_{ij}^\alpha Q^{\alpha\beta} \Lambda_{kl}^\beta . \tag{3.14}$$

The Dyson–Schwinger equation then reads

$$\eta_{ij}(x) = -Q_{ij,kl}(G_\eta)_{lk}(x,x) . \tag{3.15}$$

It is pictorially represented in Fig. 3.1. This non-linear, generally non-local equation displays strong similiarities to the gap equation of a superconductor. Also we will see in the following the solution can be interpreted in analogy to the superconductor. The vacuum solution $\langle\eta\rangle = const \neq 0$ describes a dynamically generated mass (gap at the fermi surface) and implies a non-vanishing quark condensate $\langle\bar{q}q\rangle$ (Cooper pair condensate). The mesons are the elementary small-amplitude vibrational excitation of the non-trivial ground state similar to the pairons of the superconductor. In order to determine their properties we only have to look for the eigenmodes of the fluctuations $\varphi = \eta - \langle\eta\rangle$.

$$\eta \quad = $$

Fig. 3.1. The pictorial representation of the Dyson–Schwinger equation (3.15). A full line represents the quark Green's function and Q stands for the two–body quark interaction.

Assuming that the fluctuations φ are small in amplitude it is natural to expand the quark propagator

$$G_\eta = G_{\langle\eta\rangle+\varphi}$$
$$= \left(G_{\langle\eta\rangle}^{-1}(1 - G_{\langle\eta\rangle}\varphi)\right)^{-1}$$
$$= G_{\langle\eta\rangle} + G_{\langle\eta\rangle}\varphi G_{\langle\eta\rangle} + \dots \tag{3.16}$$

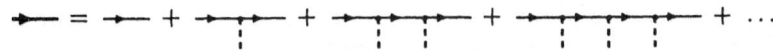

Fig. 3.2. The expansion (3.16) of the quark propagator in powers of the meson field (dashed line) φ.

This geometrical series is displayed in Fig. 3.2. Similarly, the quark loop "$\log Z_F[\eta]$" can be expanded in terms of fluctuations, see Fig. 3.3,

$$
\begin{aligned}
\log Z_F[\eta] &= \mathrm{Tr}\log G_\eta^{-1} \\
&= \mathrm{Tr}\log G_{\langle\eta\rangle}^{-1} + \mathrm{Tr}\log(1 - G_{\langle\eta\rangle}\varphi) \\
&= \mathrm{Tr}\log G_{\langle\eta\rangle}^{-1} - \sum_{n=1}^{\infty} \frac{1}{n}\,\mathrm{Tr}(G_{\langle\eta\rangle}\varphi)^n.
\end{aligned}
\tag{3.17}
$$

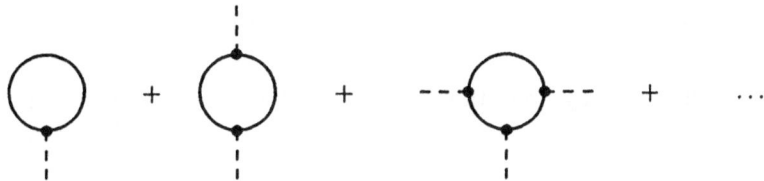

Fig. 3.3. The expansion (3.17) of the quark loop.

This expression enables us to expand the effective meson action (3.11). By construction the term linear in φ vanishes because $\langle\eta\rangle$ is the value of the η-field at the extremum (extrema) of $\mathcal{A}[\eta]$. The constant term $\mathcal{A}[\langle\eta\rangle]$ is irrelevant. The term quadratic in the fluctuations reads

$$
\begin{aligned}
\mathcal{A}^{(2)} &= -\frac{1}{2}\int d^4x\,\varphi Q^{-1}\varphi - \frac{1}{2}\mathrm{Tr}\left(\varphi G_{\langle\eta\rangle}\varphi G_{\langle\eta\rangle}\right) \\
&= \frac{1}{2}\int d^4x\, d^4y\,\varphi(x)\mathcal{D}^{-1}(x,y)\varphi(y).
\end{aligned}
\tag{3.18}
$$

The inverse meson propagator $\mathcal{D}$ is given by

$$
\mathcal{D}^{-1}(x,y) = -Q^{-1}\delta(x-y) - \Pi(x,y).
\tag{3.19}
$$

i.e. it is a sum of a local piece, the inverse of the effective quark interaction Q, and the polarization operator Π

$$
\Pi_{ij,kl}(x,y) = (G_{\langle\eta\rangle})_{jk}(x,y)(G_{\langle\eta\rangle})_{li}(y,x).
\tag{3.20}
$$

There are also terms of third and higher order in the action which we denote by $\mathcal{A}_{int}$. From eq. (3.17) one immediately sees that

$$\mathcal{A}_{int} = -\sum_{n=3}^{\infty} \frac{1}{n} \operatorname{Tr}(G_{\langle\eta\rangle}\varphi)^n \qquad (3.21)$$

This non-local expression describes the interaction of physical mesons, *e.g.* strong decays like $\phi \to 2\pi, \omega \to 3\pi$ or scattering processes like $\pi\pi \to \pi\pi$.

The equation of motion for small amplitude fluctuations

$$\frac{\partial \mathcal{A}^{(2)}}{\partial \varphi} = \mathcal{D}^{-1}\varphi = 0 \qquad (3.22)$$

is equivalent to the Bethe-Salpeter equation in ladder approximation. More explicitly it reads

$$\varphi(x) = Q\Pi(x,y)\varphi(y). \qquad (3.23)$$

This equation may be obtained from the Schwinger–Dyson equation (3.15) by

Fig. 3.4. The pictorial representation of the Bethe–Salpeter equation (3.23).

variation with respect to φ and amputating one φ line, see Figs. 3.1 and 3.4. This simply reflects the fact that the whole hierarchy of Dyson–Schwinger equations for all n-point functions may be obtained from the first one by recursive insertion and amputation.** The Bethe-Salpeter equation (3.23) is a linear integral equation which depends on the VEV $\langle\eta\rangle$ via the polarization operator $\Pi_{\langle\eta\rangle}$. For the simple NJL model this equation can be turned into an algebraic linear eigenvalue problem thereby yielding the meson masses. Before discussing the fluctuations of the meson field it is necessary to determine its VEV $\langle\eta\rangle$ using the Dyson–Schwinger equation (3.12).

** In the pseudoscalar channel the Bethe-Salpeter equation may be also obtained by an infinitesimal chiral transformation from the Dyson–Schwinger equation. This ensures the appearance of the pseudoscalar mesons as Goldstone bosons.

3.3 Dynamical Breaking of Chiral Symmetry

In order to discuss the Dyson–Schwinger equation (3.12) in more detail we decompose the generic meson field η according to its properties under Lorentz transformation

$$\eta = S + i\gamma_5 P - i\slashed{V} - i\slashed{A}\gamma_5. \tag{3.24}$$

S is a scalar, P a pseudoscalar, V a vector and A an axialvector field. All these fields are flavor matrices, $S = S^a(\lambda^a/2)_F$ and so on. Note that due to the structure of the interaction (2.24) and the Fierz transformation property (2.49) no tensor meson field appears.

Using (3.24) the chirally invariant interaction term of the NJL model in (3.4) can be rewritten as

$$\frac{1}{2}\eta Q^{-1}\eta = \frac{1}{2G_1}\mathrm{tr}(S^2 + P^2) + \frac{1}{2G_2}\mathrm{tr}(V_\mu V^\mu + A_\mu A^\mu) \tag{3.25}$$

where $G_2 = \frac{1}{2}G_1 = \frac{1}{3}g^2\kappa$. Performing again the shift in the integration variable $\eta \to \eta - m_0$, $i.e.$ $S \to S - m_0$, the generating functional Z_{NJL} (3.11) reads

$$Z_{NJL} = \int \mathcal{D}S\,\mathcal{D}P\,\mathcal{D}V\,\mathcal{D}A \; e^{\mathcal{A}[S,P,V,A]}$$

$$\mathcal{A}[S,P,V,A] = -\frac{1}{2G_1}\int d^4x \; \mathrm{tr}\big((S - m_0)^2 + P^2\big)$$

$$- \frac{1}{2G_2}\int d^4x \; \mathrm{tr}\big(V_\mu V^\mu + A_\mu A^\mu\big)$$

$$+ \mathrm{Tr}\log\Big(i(\slashed{\partial} + \slashed{V} + \slashed{A}\gamma_5) - (S + i\gamma_5 P)\Big) . \tag{3.26}$$

In order to keep the discussion of chiral symmetry breaking as pedagogical as possible we will make some further simplifications. First, we set $V_\mu = A_\mu = 0$ which is an often used form of the NJL model. Second, we consider only one flavor, $N_f = 1$. Then only one scalar and one pseudoscalar meson exist. Instead of two real fields we use one complex field $M = S + iP$. It is advantageous to introduce the angular decomposition, $i.e.$ polar coordinates,

$$M = \phi U = \phi e^{i\Theta} = \phi(\cos\Theta + i\sin\Theta) \tag{3.27}$$

where $\phi \in [0, \infty]$ is the chiral radius field, $U \in U(1)$ is usually called chiral field and $\Theta \in [0, 2\pi]$ chiral angle. Using the chiral projectors $P_{R,L}$ (2.30) one easily shows that

$$S + i\gamma_5 P = (S + iP)P_R + (S - iP)P_L$$

$$= M P_R + M^\dagger P_L$$

$$= \phi e^{i\gamma_5\Theta} . \tag{3.28}$$

For the considered simplified case the effective action (3.26) reads

$$A[S, P] = A[\phi, \Theta]$$
$$= -\frac{1}{2G_1} \int d^4x (\phi^2 - 2\phi m_0 \cos \Theta + m_0^2) + \text{Tr} \log(i\not{\partial} - \phi e^{i\gamma_5 \Theta}) .$$

$$(3.29)$$

Let us make for the moment the calculations even easier. This can be achieved by using the so–called strong coupling approximation where the gradients in the fermion determinant are neglected. This amounts to the replacement

$$\text{Tr} \log(i\not{\partial} - \phi e^{i\gamma_5 \Theta}) \simeq \omega \text{Tr} \log(-\phi e^{i\gamma_5 \Theta}/\omega^{1/4}) \qquad (3.30)$$

where the scale $\omega^{1/4}$ was introduced because the functional determinant of a function only (no derivatives) is diverging. The constant ω is meant to absorb these divergencies and to regularize the above expression in a suitable way. The r.h.s. of (3.30) can be simplified further:

$$\omega \text{Tr} \log(-\phi e^{i\gamma_5 \Theta}/\omega^{1/4}) = \omega \int d^4x (\text{tr} \log(-\phi/\omega^{1/4}) + i\Theta \text{tr}\gamma_5)$$

$$= \omega \int d^4x \log \det(-\phi/\omega^{1/4})$$

$$= 4N_c\omega \int d^4x \log |\phi/\omega^{1/4}|. \qquad (3.31)$$

As a consequence of the global chiral invariance the chiral angle field Θ has disappeared from the quark determinant. Unfortunately, the strong coupling approximation makes the quark determinant also invariant with respect to local chiral transformations which is, of course, unphysical.***

Using (3.30) and (3.31) the corresponding simplified effective action is given by

$$A[\phi, \Theta] = -\int d^4x \left(\frac{1}{2G_1}(\phi^2 - 2\phi m_0 \cos \Theta + m_0^2) - 4N_c\omega \log |\phi/\omega^{1/4}| \right) .$$

$$(3.32)$$

The expression in parentheses is exactly the effective potential $\mathcal{V}[\phi, \Theta]$ defined by

$$A[\phi, \Theta] = -\int d^4x \mathcal{V}[\phi, \Theta]. \qquad (3.33)$$

It is shown in Fig. 3.5 for $m_0 = 0$ as a function of ϕ. Note that for $m_0 = 0$ it is independent of Θ. One clearly sees that the minimum of the potential is at a non-vanishing value of $\langle \phi \rangle$ which may be interpreted as a dynamically generated fermion mass, the constituent quark mass. This term spoils the chiral invariance,

---

*** The alert reader has probably already noticed that the chiral anomaly escapes in the strong coupling limit.

i.e. independence of Θ, of the ground state: chiral symmetry is spontaneously broken in a dynamical fashion. Fluctuations along the 'valley', *i.e.* the direction of Θ, cost no energy; therefore these fluctuations correspond to a massless particle, the Goldstone boson. For $m_0 \neq 0$ the effective potential is not independent of Θ. Fluctuations out of the minimum feel a restoring force and the Θ-field becomes massive. This is the way the pion mass is induced by the current masses of quarks.

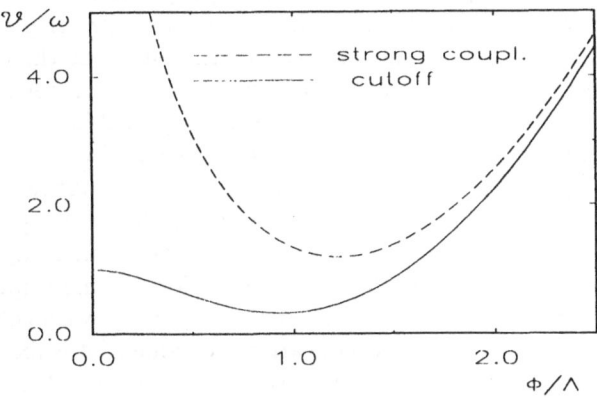

Fig. 3.5. The effective potential in the strong coupling limit (dashed line) (3.33) with (3.32) and with momenta included (3.42) (solid line) as a function of the value of the scalar field ϕ.

Varying $\mathcal{A}$ (3.32) with respect to ϕ and Θ yields the Schwinger–Dyson equation in the strong coupling limit

$$\phi = m_0 \cos \Theta + 4N_c \frac{\omega G_1}{\phi}$$

$$\phi m_0 \sin \Theta = 0 . \tag{3.34}$$

One sees immediately that in the chiral limit ($m_0 = 0$) Θ is arbitrary which reflects the underlying chiral symmetry. For $m_0 \neq 0$ one has $\Theta = 0 (\mathrm{mod} 2\pi)$ or $\Theta = \pi (\mathrm{mod} 2\pi)$. The last case corresponds to a maximum of the action (3.32), so we consider only the first case. For the VEV of the field ϕ one obtains

$$\langle \phi \rangle = \frac{1}{2} m_0 \pm \sqrt{(m_0/2)^2 + 4N_c \omega G_1} . \tag{3.35}$$

As already stated this field expectation value may be interpreted as a dynamically generated constituent quark mass. Only the upper sign corresponds to a physical solution as we will see in the following. Also the divergence of the effective potential for $\phi \to 0$ is an artifact of the strong coupling approximation which breaks down for small ϕ as become obvious from eq. (3.30). Nevertheless, eqs. (3.34) already display the generic character of Dyson–Schwinger equations. Especially, they will be independent of Θ in the chiral limit $m_0 = 0$.

Keeping the gradients in the fermion determinant in the effective action (3.29) one still needs some regularization procedure because the effective interaction (2.24) is non–renormalizable. The regularization should respect the symmetries of the classical Lagrangians as much as possible. So $e.g.$ Lorentz invariance is kept if one uses an $O(4)$ invariant momentum cutoff in the Euclidean formulation of the model. As we will couple the mesons to external gauge fields ($e.g.$ the photon) it will prove to be convenient to use regularization schemes that respect gauge invariance. Such methods are $e.g.$ Schwinger's proper time method [Sch51], dimensional regularization [PT84, IZ85] or the Paul–Villars prescription [PT84, IZ85]. In a renormalizable theory one can remove the cutoff at the end of the calculation, and the result for physical observables should not depend on the used regularization scheme. However, for the non–renormalizable QFD (2.24) the cutoff has to stay finite. It reflects the physical scale where QCD changes from a theory with strong (non–perturbative) interaction to the perturbative regime. The cutoff Λ characterizes the region $p^2 < \Lambda^2$ where the model is applicable. It may be therefore fitted to some physical observable, $e.g.$ the pion decay constant. The model is only defined together with the regularization scheme employed, and physical results will in general depend on it. The hope is that physical observables will depend only slightly on the used scheme after the value of Λ is fixed in order to reproduce a certain quantity. From the discussion in Chap. 2 it is obvious that QCD may provide a different suppression of high-momentum contributions to hadronic properties for every different quantity considered.

In the following we will use Schwinger's proper time regularization [Sch51]. The underlying idea is that for a positive definite Hermitean operator A the logarithm may be written as

$$\log A = - \lim_{\Lambda \to \infty} \int_{1/\Lambda^2}^{\infty} \frac{d\tau}{\tau} e^{-\tau A} + const. \tag{3.36}$$

We will make this substituion for the logarithm in eq. (3.29), however, we will keep the cutoff Λ finite.

Since we are interested here in the vacuum configuration we may restrict ourselves to constant fields without involving any further approximation. In this case the logarithm in eq. (3.29) may be rewritten as

$$\mathrm{Tr} \log(i\slashed{\partial} - \phi e^{i\gamma_5 \Theta}) = \frac{1}{2} \mathrm{Tr} \log(\partial^2 + \phi^2). \tag{3.37}$$

This term is given after the substitution (3.36) by

$$\mathcal{A}_F := -\frac{1}{2} \mathrm{Tr} \int_{1/\Lambda^2}^{\infty} \frac{d\tau}{\tau} e^{-\tau(\partial^2 + \phi^2)}. \tag{3.38}$$

We will evaluate the functional trace using position and momentum eigenstates

$$
\begin{aligned}
\mathcal{A}_F &= -\frac{1}{2}\mathrm{tr}_c\mathrm{tr}_D \int d^4x \langle x| \int \frac{d\tau}{\tau} e^{-\tau(\partial^2+\phi^2)}|x\rangle \\
&= -\frac{1}{2}N_c\mathrm{tr}_D \int d^4x \int d^4q \int d^4k \langle x|q\rangle\langle q| \int \frac{d\tau}{\tau} e^{-\tau(\partial^2+\phi^2)}|k\rangle\langle k|x\rangle \\
&= -\frac{1}{2}4N_c \int d^4x \int \frac{d^4k}{(2\pi)^4} \int \frac{d\tau}{\tau} e^{-\tau(-k^2+\phi^2)}
\end{aligned}
\tag{3.39}
$$

where tr_c and tr_D denotes the trace over color and Dirac indices, respectively. We have also used that $\langle k|x\rangle = e^{-ikx}/(2\pi)^2$. Therefore, $\langle x|k\rangle\langle k|x\rangle = 1/(2\pi)^4$ is only a constant. Note that we use negative Euclidean metric and that $-k^2 = -k_\mu k^\mu = -g^{\mu\nu}k_\mu k_\nu = \delta^{\mu\nu}k_\mu k_\nu = |k|^2$. The integral over momenta can be easily done in spherical coordinates

$$
\begin{aligned}
\int \frac{d^4k}{(2\pi)^4} e^{-\tau(-k^2+\phi^2)} &= \frac{1}{16\pi^4} \int d|k||k|^3 \int_{S^3} d\Omega e^{-\tau(|k|^2+\phi^2)} \\
&= \frac{1}{16\pi^2} \int dz z e^{-\tau(z+\phi^2)} \\
&= \frac{1}{16\pi^2\tau^2} e^{-\tau\phi^2}
\end{aligned}
\tag{3.40}
$$

where we have used $\int_{S^3} d\Omega = 2\pi^2$. Using eqs. (3.40) and (3.39) as well as the definitions of the incomplete Γ functions we obtain[†]

$$
\begin{aligned}
\mathcal{A}_F &= -\frac{1}{2}\frac{N_c}{4\pi^2} \int d^4x \phi^4 \Gamma(-2, \frac{\phi^2}{\Lambda^2}) \\
&= -\frac{N_c}{16\pi^2} \int d^4x \left(\phi^4\Gamma(0, \frac{\phi^2}{\Lambda^2}) - \phi^2\Lambda^2 e^{-\phi^2/\Lambda^2} + \Lambda^4 e^{-\phi^2/\Lambda^2} \right).
\end{aligned}
\tag{3.41}
$$

The effective potential in the chiral limit is given by

$$
\mathcal{V}[\phi] = \frac{1}{2G_1}\phi^2 + \frac{N_c}{16\pi^2} \left(\phi^4\Gamma(0, \frac{\phi^2}{\Lambda^2}) - (\phi^2 - \Lambda^2)\Lambda^2 e^{-\phi^2/\Lambda^2} \right).
\tag{3.42}
$$

Using that $\omega = \Lambda^4/32\pi^2$ for a sharp Euclidean cutoff the analogous expression in the strong coupling limit is given by (using eqs. (3.32) and (3.33))

$$
\mathcal{V}[\phi] = \frac{1}{2G_1}\phi^2 - N_c\frac{\Lambda^4}{16\pi^2} \log\frac{\phi^2}{\Lambda^2}
\tag{3.43}
$$

[†] The functions

$$
\Gamma(u, x) = \int_x^\infty d\tau \tau^{u-1} e^{-\tau}
$$

are known as incomplete Γ-functions. Especially,

$$
\Gamma(0, x) = -\log x - \gamma + \mathcal{O}(x) \quad \text{for} \quad x \to 0
$$

where $\gamma = 0.57721$ is Euler's constant, see e.g. Chaps. 5 and 6 of [AS65] for more details.

28

where we added a constant term proportional to $\log 32\pi^2$ to make the expression (3.43) identical to (3.42) for large amplitudes of the field ϕ where the strong coupling approximation is valid. In the limit $\phi \to 0$ this expression diverges whereas the potential in proper–time regularization (3.42) behaves like

$$\mathcal{V}[\phi] \to \frac{1}{2G_1}\phi^2 - \frac{N_c}{16\pi^2}\left(\phi^4 \log \frac{\phi^2}{\Lambda^2} + \Lambda^4\right) + \ldots \qquad (3.44)$$

and is therefore regular in contrast to the logarithm in eq. (3.43), see Fig. 3.5.

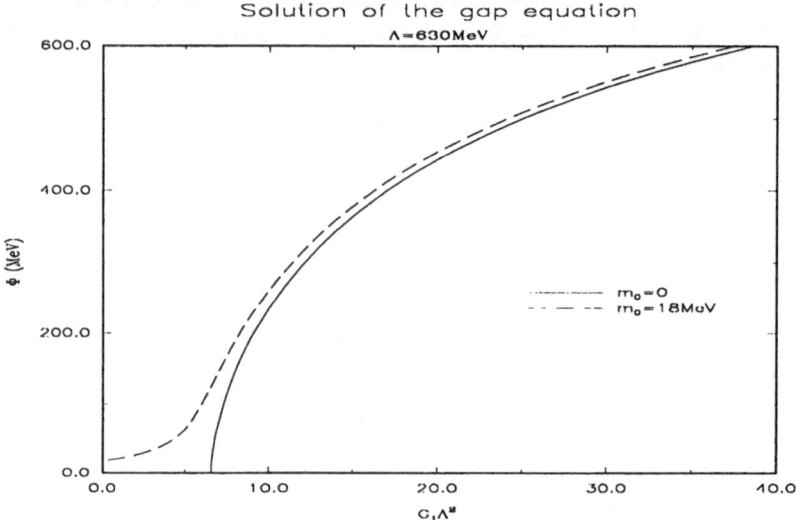

Fig. 3.6. The solution of the gap equation (3.45) for vanishing current mass $m^0 = 0$ (solid line) and $m^0 = 18\text{MeV}$ (dashed line).

For completeness we quote here the Dyson–Schwinger equation in proper time regularization as compared to eqs. (3.34) in the strong coupling approximation. Obviously, the equation $\phi m_0 \sin \Theta = 0$ will not change. Choosing now $\Theta = 0$ the equation for the VEV of ϕ reads

$$\langle\phi\rangle = m_0 + 2G_1\frac{N_c}{4\pi^2}\langle\phi\rangle^3\Gamma(-1,\frac{\langle\phi\rangle^2}{\Lambda^2}). \qquad (3.45)$$

The quantity

$$\langle\bar{q}q\rangle = -\frac{N_c}{4\pi^2}\langle\phi\rangle^3\Gamma(-1,\frac{\langle\phi\rangle^2}{\Lambda^2}) \qquad (3.46)$$

is usually called the quark condensate and provides a measure for the dynamical breaking of chiral symmetry.

In Fig. 3.6 the solution of eq. (3.45) is shown for $\Lambda = 630$MeV. In the chiral limit, $m_0 = 0$, there is always the trivial solution $\langle\phi\rangle = 0$. A non–trivial solution starts to exist only if the coupling constant G_1 exceeds a critical value. For finite current masses the trivial solution ceases to exist.

3.4 The Bethe–Salpeter Equation for Pseudoscalar Mesons

As we have seen in the preceding section the pseudoscalar mesons are due to the dynamical breaking of chiral symmetry of a special nature: they are Goldstone bosons related to this symmetry. Therefore we will discuss their Bethe–Salpeter equation (see eqs. (3.22,3.23)) in more detail. Additionally, we relax our simplification of using only one flavor. Considering two light flavors of equal mass their Schwinger–Dyson equation is given by

$$\langle\phi_{ij}\rangle = \delta_{ij} m_i$$
$$m_i = m_i^0 - 2G_1\langle\bar{q}q\rangle_i$$
$$\langle\bar{q}q\rangle_i = -m_i^3 \frac{N_c}{4\pi^2}\Gamma(-1, m_i^2/\Lambda^2). \tag{3.47}$$

The dynamically generated constituent quark masses m_i, $i = u, d$, (and also the quark condensates $\langle\bar{q}q\rangle_i$) are equal, $m_u = m_d =: m$ and $\langle\bar{u}u\rangle = \langle\bar{d}d\rangle$, respectively, if the quark current masses are equal, $m_u^0 = m_d^0 =: m_0$. In this section we will restrict ourselves to this isospin symmetric limit.

The physical meson excitations are given by small amplitude fluctuations around the translational invariant vacuum field configuration (3.47). As the fluctuations are in general functions of the space–time coordinate x_μ we have to be much more careful in the evaluation of the fermion determinant as we have been so far. Expanding the effective action

$$\mathcal{A} = \mathcal{A}_m + \mathcal{A}_F$$
$$\mathcal{A}_m = \int d^4x \left(-\frac{1}{4G_1}\text{tr}(M^\dagger M - m_0(M + M^\dagger) + m_0^2)\right)$$
$$\mathcal{A}_F = -\frac{1}{2}\int_{1/\Lambda^2}^{\infty} \frac{ds}{s}\text{Tr}\exp\left(-s\not{p}^\dagger\not{p}\right)$$
$$\not{p}^\dagger\not{p} = \partial^2 + [i\not{\partial}, P_R M + P_L M^\dagger] + P_R M^\dagger M + P_L M M^\dagger. \tag{3.48}$$

up to second order in the fluctuations of the pseudoscalar field $\Theta(x)$ allows to extract the inverse propagator for the pseudoscalar mesons. Since we are not interested in the fluctuations of the scalar field we use the VEV $\phi = m\mathbf{1}$. The complex field M is then given by

$$M = S + iP = mU = me^{i\Theta(x)}. \tag{3.49}$$

Obviously, the VEV of the pseudoscalar field $\Theta(x)$ is zero. The field U is expanded for small–amplitude fluctuations of the pseudoscalar meson field as

$$U = e^{i\Theta} =: e^{2i\pi} = 1 + 2i\pi - 2\pi^2 + \ldots. \tag{3.50}$$

The factor 2 is hereby introduced for later convenience. Using $\pi = \pi^i(\tau^i/2)$ the fields π describe the pions.

We have to expand the action up to second order in the fields π. As the stationary point of the action occurs for $\pi = 0$ there is no linear term. First, we will consider the mesonic mass term $\mathcal{A}_m$

$$\mathcal{A}_m = -\frac{1}{2G_1} \int d^4x \left((m - m_0)^2 + 4m_0 m \pi^2\right) + \mathcal{O}(\pi^3). \qquad (3.51)$$

Introducing the Fourier transform of the fluctuating angle, $\pi(q)$, the bilinear term of (3.51) is obtained to be:

$$\int \frac{d^4q}{(2\pi)^4} \sum_{i=1}^{3} \frac{1}{2} \pi^i(-q) \pi^i(q) \frac{m_0 m}{G_1}. \qquad (3.52)$$

In order to expand the fermion determinant $\mathcal{A}_F$ we rewrite the operator

$$\displaystyle{\not{D}^\dagger \not{D}} = A_0 + A_1 + A_2 + \ldots \qquad (3.53)$$

where A_k is of k-th order in the field π. Using eqs. (3.48) and (3.50) we obtain

$$
\begin{aligned}
A_0 &= \partial^2 + m^2 \\
A_1 &= 2\gamma_5 m(\not{\partial}\pi) \\
A_2 &= 0.
\end{aligned}
\qquad (3.54)
$$

Obviously only derivatives of π can occur which can be understood from the chiral invariance of $\mathrm{Det}\,\not{D}^\dagger \not{D}$. Using now

$$
\begin{aligned}
\mathcal{A}_F &= -\frac{1}{2} \int_{1/\Lambda^2}^{\infty} \frac{ds}{s} \mathrm{Tr}\exp\left(-s\not{D}^\dagger \not{D}\right) \\
&= -\frac{1}{2} \int_{1/\Lambda^2}^{\infty} \frac{ds}{s} \mathrm{Tr}e^{-sA_0} + \frac{1}{2}\int_{1/\Lambda^2}^{\infty} ds \int_0^1 d\zeta \mathrm{Tr}e^{-s\zeta A_0} A_2 e^{-s(1-\zeta)A_0} \\
&\quad -\frac{1}{2}\int_{1/\Lambda^2}^{\infty} ds\, s \int_0^1 d\zeta \int_0^{1-\zeta} d\eta\, e^{-s\eta A_0} A_1 e^{-s(1-\zeta-\eta)A_0} A_1 e^{-s\zeta A_0} + \mathcal{O}(\pi^3)
\end{aligned}
$$

$$(3.55)$$

we can systematically expand this part of the action. The flavor (or isospin) trace only gives an overall factor 2, the color trace a factor N_c and the Dirac trace a factor 4. The functional trace will be performed using momentum eigenstates $\pi(q)$. As $A_2 = 0$ in the isospin limit we have to calculate the following parameter integral

$$\int_0^1 d\zeta \int_0^{1-\zeta} d\eta \left(e^{-s(\zeta+\eta)(k^2+m^2)} e^{-s(1-\zeta-\eta)((k+q)^2+m^2)} \right.$$
$$\left. + e^{-s(\zeta+\eta)(k^2+m^2)} e^{-s(1-\zeta-\eta)((k+q)^2+m^2)} \right)$$
$$= \frac{1}{2} \int_0^1 d\alpha \int_{-\alpha}^{\alpha} d\beta \left(e^{-s\alpha(k^2+m^2)} e^{-s(1-\alpha)((k+q)^2+m^2)} \right.$$
$$\left. + e^{-s\alpha(k^2+m^2)} e^{-s(1-\alpha)((k+q)^2+m^2)} \right)$$
$$= \int_0^1 d\alpha e^{-s\alpha(k^2+m^2)} e^{-s(1-\alpha)((k+q)^2+m^2)}. \tag{3.56}$$

As the momentum k only appears in this exponential we may shift $k \to k - (1-\alpha)q$ in the second term without changing the value of the integral thereby obtaining

$$\int \frac{d^4k}{(2\pi)^4} \int_0^1 d\alpha e^{-s((k+(1-\alpha)q)^2 + \alpha(1-\alpha)q^2 + m^2)}$$
$$= \int \frac{d^4k}{(2\pi)^4} \int_0^1 d\alpha e^{-s(k^2 + \alpha(1-\alpha)q^2 + m^2)}. \tag{3.57}$$

Using

$$4N_c \int \frac{d^4k}{(2\pi)^4} e^{-sk^2} = \frac{N_c}{4\pi^2} \tag{3.58}$$

as well as the definition of the incomplete Γ–function, the term bilinear in pseudoscalar fields arising from $\mathcal{A}_F$ reads

$$\int \frac{d^4q}{(2\pi)^4} \sum_{i=1}^{3} \frac{1}{2} \pi^i(-q)\pi^i(q)(-q^2)m^2 \frac{N_c}{4\pi^2} \int_0^1 d\alpha \Gamma(0, (m^2 + \alpha(1-\alpha)q^2)/\Lambda^2). \tag{3.59}$$

Comparing with eq. (3.18) allows us to extract the inverse pion propagator

$$D^{-1}(q^2) = -\frac{2m_0 m}{G_1} - \Pi(q^2) \tag{3.60}$$

where the polarization operator $\Pi(q^2)$ is given by

$$\Pi(q^2) = -2q^2 f_\pi^2(q^2)$$
$$f_\pi^2(q^2) = m^2 \frac{N_c}{4\pi^2} \int_0^1 d\alpha \, \Gamma\left(0, [m^2 + \alpha(1-\alpha)q^2]/\Lambda^2\right). \tag{3.61}$$

The Bethe–Salpeter equation which determines the physical meson masses m_π is equivalent to the condition that the meson propagator has a pole:

$$D^{-1}(q^2 = -m_\pi^2) = 0. \tag{3.62}$$

Note that $f_\pi^2(q^2 = -m_\pi^2)$ is then the corresponding meson decay constant. We want to emphasize here that the Bethe–Salpeter equation (3.60) – (3.62) is the one in ladder approximation and that no other approximations as e.g. gradient or heat kernel expansions have been made. This also implies that the decay constants are evaluated on the corresponding meson mass shell. The pion decay constant is then given by

$$f_\pi^2 = m^2 \frac{N_c}{4\pi^2} \int_0^1 d\alpha\, \Gamma\Big(0, [m^2 - \alpha(1-\alpha))m_\pi^2]/\Lambda^2\Big). \tag{3.63}$$

One clearly sees that in the chiral limit ($m_\pi = 0$) the expression

$$f_\pi^2 = m^2 \frac{N_c}{4\pi^2} \Gamma(0, (m/\Lambda)^2) \tag{3.64}$$

calculated by means of a derivative expansion becomes exact. Using eqs. (3.60) and (3.61) it is trivial to show that in the chiral limit $m_u^0 = 0$ the Bethe–Salpeter equation (3.62) is solved for $q^2 = -m_\pi^2 = 0$. Neglecting the momentum dependence of f_π^2 in eq. (3.61) one obtains

$$m_\pi^2 f_\pi^2 = \frac{m_0 m}{G_1} = 2m_0 \langle \bar{u}u \rangle \frac{m}{m - m_0} \approx 2m_0 \langle \bar{u}u \rangle \tag{3.65}$$

where we used the gap equation (3.47) to express the coupling constant G_1 in terms of the quark condensates. The approximate relation $\frac{1}{2}m_\pi^2 f_\pi^2 = m_0 \langle \bar{u}u \rangle$ is known as Gell-Mann–Renner–Oakes relation [GMOR68], see also Chap. 5 of [CL88]. The Bethe–Salpeter equation for three flavors and unequal masses is treated in Appendix A of [WAR92].

3.5 The Gauged Linear σ Model

In this section we will proceed with our analysis of the effective meson theory defined by eq. (3.11). However, this time we will keep the vector and axialvector meson fields. In order to be more general we will also keep the number of flavors arbitrary and treat the fields as non–commuting flavor matrices. Using the chiral projectors P_L and P_R we rewrite the Dirac operator in the fermion determinant as

$$\begin{aligned}
i\slashed{D} &= i\slashed{\partial} - \eta \\
&= i(\slashed{\partial} + \slashed{V} + \slashed{A}\gamma_5) - (S + i\gamma_5 P) \\
&= i(\slashed{\partial} + \slashed{V}_R)P_R + i(\slashed{\partial} + \slashed{V}_L)P_L - (M P_R + M^\dagger P_L)
\end{aligned} \tag{3.66}$$

where the left (right) handed vector fields are given by

$$V_{L,R}^\mu = V^\mu \mp A^\mu. \tag{3.67}$$

In the effective theory we want to have the Dirac operator (3.66) as simple as possible. Therefore we eliminate the chiral field U appearing in the angular decomposition $M = U\phi$ from the fermion determinant by a chiral rotation. As a first step we rewrite the fermion determinant again in terms of the functional integral over Grassmann variables, *i.e.* quark fields,

$$\mathcal{A}_F[\eta] = \mathrm{Tr}\log(i\not{D}) = \log\int \mathcal{D}q\mathcal{D}\bar{q}\exp\left(\int \bar{q}i\not{D}q\right). \tag{3.68}$$

On these fields we apply a chiral rotation. For definiteness we choose a chiral left rotation

$$\begin{aligned} q_L &= Uq'_L, & q_R &= q'_R, \\ \bar{q}_L &= \bar{q}'_L U^\dagger, & \bar{q}_R &= \bar{q}'_R, \end{aligned} \tag{3.69}$$

or in a more compact notation

$$\begin{aligned} q &= \Omega_L q', & \Omega_L &= P_R + P_L U, \\ \bar{q} &= \bar{q}'\Omega_R^\dagger, & \Omega_R &= P_L + P_R U. \end{aligned} \tag{3.70}$$

Requiring now $\bar{q}i\not{D}q = \bar{q}'(i\not{D}')q'$ amounts to $i\not{D}' = \Omega_R^\dagger i\not{D}\Omega_L$ and therefore

$$i\not{D}' = i(\not{\partial} + \not{V}'_R)P_R + i(\not{\partial} + \not{V}'_L)P_L - \phi \tag{3.71}$$

where

$$\begin{aligned} V'^\mu_R &= V^\mu_R \\ V'^\mu_L &= U^\dagger V^\mu_L U + L^\mu, & L^\mu &= U^\dagger \partial^\mu U. \end{aligned} \tag{3.72}$$

Note that in eq. (3.71) only the scalar part ϕ of $M = U\phi$ survives. The vector field L^μ appearing in (3.72) is the left current of the chiral field. Note also that the transformation (3.69) changed the pseudoscalar coupling of the chiral field into a derivative coupling. Using eq. (3.69) it is easy to prove that a constant chiral field does not contribute to the fermion determinant because it can be removed by a chiral rotation. This explains why only derivatives of the pion field occur in eqs. (3.53) and (3.54), or why the fermion determinant becomes independent of the chiral field in the strong coupling approximation.

Note that the general chiral transformation

$$\begin{aligned} q &\to \Omega_1 q \\ \bar{q} &\to \bar{q}\Omega_2 \\ \Omega_1 &= P_L\xi_L + P_R\xi_R \\ \Omega_2 &= \xi_L^\dagger P_R + \xi_R^\dagger P_L \end{aligned} \tag{3.73}$$

also removes the chiral field from the pseudoscalar sector of the fermion determinant as long as the $\xi_{L,R}$ fulfill

$$U = \xi_L^\dagger \xi_R. \tag{3.74}$$

We recover the above discussed case setting $\xi_R = 1$ and a symmetric formulation called the unitary gauge for $\xi_L^\dagger = \xi_R$.

Let us mention already here that a general chiral transformation does not leave the integration measure of the functional integral **not** invariant but generates a field dependent phase in the action. Defining the Jacobian $J(U)$ of the chiral transformation by

$$\int \mathcal{D}q\mathcal{D}\bar{q} \exp\left(\int \bar{q}i\!\!\not{D}q\right) = J(U) \int \mathcal{D}q'\mathcal{D}\bar{q}' \exp\left(\int \bar{q}'i\!\!\not{D}'q'\right) \quad (3.75)$$

the fermion determinant can be written in two equivalent ways

$$\mathcal{A}_F[\eta] = \text{Tr}\log i\!\!\not{D}(\phi,U,V,A)$$
$$= \log J(U) + \text{Tr}\log i\!\!\not{D}'(\phi,V',A'). \quad (3.76)$$

In general the Jacobian $J(U) \neq 1$, see Sect. 3.7. The absolute value of J is, however, unity so that in Euclidean space $\mathcal{A}_{\text{WZW}} := \log J$ is imaginary. $\mathcal{A}_{\text{WZW}}$ is the so-called Wess–Zumino–Witten term and may be regarded as the field theoretical analogy of the Berry phase.

Introducing V'^{μ} and A'^{μ} defined by

$$V'^{\mu}_{R,L} = V'^{\mu} \pm A'^{\mu} \quad (3.77)$$

as new integration variables it is straightforward to show that the integration measure is invariant, $\mathcal{D}V\mathcal{D}A = \mathcal{D}V'\mathcal{D}A'$. However, the mass term changes

$$\text{tr}(V^2 + A^2) = \frac{1}{2}\text{tr}(V_L^2 + V_R^2)$$
$$= \text{tr}((V'_{\mu} - v_{\mu})^2 + (A'_{\mu} - a_{\mu})^2) \quad (3.78)$$

where $v_{\mu} = \frac{1}{2}L_{\mu}$ and $a_{\mu} = -\frac{1}{2}L_{\mu}$ in the case of a chiral left transformation (3.69), (3.70). Omitting the primes the generating functional then reads

$$Z = \int \mathcal{D}\phi\, \mathcal{D}U\, \mathcal{D}V_{\mu}\, \mathcal{D}A_{\mu}\, e^{\mathcal{A}_{\text{eff}}(\phi,U,V,A)}$$

$$\mathcal{A}_{\text{eff}}(\phi,U,V,A) = \mathcal{A}_m + \log J(U) + \mathcal{A}_F(\phi,V,A)$$

$$\mathcal{A}_m = -\frac{1}{4G_1} \int d^4x\, \text{tr}(\phi^2 - (m_0 U\phi + U^+\phi m_0) + m_0^2)$$

$$-\frac{1}{4G_2} \int d^4x\, \text{tr}((V-v)^2 + (A-a)^2)$$

$$\mathcal{A}_F = \text{Tr}\log(i(\not{\partial} + \not{V} + \not{A}\gamma_5) - \phi). \quad (3.79)$$

Including vector fields and/or more than two flavors the quark loop $\mathcal{A}_F$ is in general complex in the Euclidean formulation. Therefore we separate real and imaginary parts[‡]

$$\mathcal{A}_F = \text{Tr}\log(i\!\!\not{D})$$
$$= \frac{1}{2}\text{Tr}\log(\not{D}^\dagger \not{D}) + \frac{1}{2}\text{Tr}\log((\not{D}^\dagger)^{-1}\not{D})$$
$$=: \mathcal{A}_F^R + \mathcal{A}_F^I. \quad (3.80)$$

[‡] This operation makes use of the relation $\log AB = \log A + \log B$. One should emphasize, however, that this relation does no longer strictly hold as soon as the fermion determinant is regularized.

The real part $\mathcal{A}_F^R$ is a divergent quantity and a finite cutoff is needed whereas the imaginary part $\mathcal{A}_F^I$ is convergent. We will discuss the real part of the action postponing the discussion of the imaginary part to the end of this chapter. The purpose of the remainder of this chapter will be to display the low–energy meson physics contained in the action (3.79). As we are interested in the low–energy limit it is sufficient to treat the real part of the quark loop $\mathcal{A}_F^R$ in a gradient expansion. The "heat kernel method", a covariant gradient expansion, is discussed in Appendix A.

In order to compactify notation we introduce the following definitions

$$d_\mu := \partial_\mu + \Gamma_\mu \quad \text{where} \quad \Gamma_\mu = V_\mu + A_\mu \gamma_5$$

$$a := i\slashed{\Delta}\phi + \phi^2 + \frac{1}{4}[\gamma^\mu, \gamma^\nu]\Gamma_{\mu\nu} - \mu^2$$

$$\text{where} \quad \Delta_\mu \phi = \partial_\mu \phi + [V_\mu, \phi] - \{A_\mu, \phi\}\gamma_5, \quad \Gamma_{\mu\nu} = [d_\mu, d_\nu]. \quad (3.81)$$

μ is an arbitrary energy scale which will be later identified with the constituent quark mass. Note also the hermiticity properties of the quantities

$$\phi = \phi^\dagger, \quad V_\mu = -V_\mu^\dagger, \quad A_\mu = -A_\mu^\dagger, \quad \gamma_\mu = -\gamma_\mu^\dagger. \quad (3.82)$$

With the definitions (3.81) and using (3.82) the operator $\slashed{D}^\dagger \slashed{D}$ is simply given by

$$\slashed{D}^\dagger \slashed{D} = d_\mu d^\mu + a + \mu^2. \quad (3.83)$$

The real part of the quark loop $\mathcal{A}_F^R$ (3.80) containing this operator is given in second order heat kernel expansion [ER86] (see Appendix A)

$$\mathcal{A}_F^R = \frac{1}{g_V^2} \int d^4x \, \mathrm{tr} \left(\frac{3}{2} (\mathcal{D}_\mu \phi \mathcal{D}^\mu \phi - \{A_\mu, \phi\}\{\phi, A^\mu\}) + \frac{1}{2} \left(F_{\mu\nu}^V F_V^{\mu\nu} + F_{\mu\nu}^A F_A^{\mu\nu} \right) \right)$$
$$-\mathcal{V}_F[\phi] \quad (3.84)$$

where $\mathcal{V}_F[\phi]$ is the potential coming from the quark loop, g_V is a regularization dependent parameter and

$$\mathcal{D}_\mu \phi = \partial_\mu \phi + [V_\mu, \phi],$$
$$F_{\mu\nu}^V = \partial_\mu V_\nu - \partial_\nu V_\mu + [V_\mu, V_\nu] + [A_\mu, A_\nu],$$
$$F_{\mu\nu}^A = \partial_\mu A_\nu - \partial_\nu A_\mu + [V_\mu, A_\nu] + [A_\mu, V_\nu]. \quad (3.85)$$

Let us return for the following discussion to the isospin limit, $N_f = 2, m_0 = m_0^u = m_0^d$. g_V is then given by

$$g_V = \left(\frac{2}{3} \frac{N_c}{16\pi^2} \Gamma(0, \frac{m^2}{\Lambda^2}) \right)^{-\frac{1}{2}} \quad (3.86)$$

and the potential $\mathcal{V}_F$ by

$$\mathcal{V}_F[\phi] = \frac{N_c}{16\pi^2} \left(\Gamma(0, \frac{m^2}{\Lambda^2}) \mathrm{tr}(\phi^4) - 2\Lambda^2 \mathrm{tr}(\phi^2) e^{-m^2/\Lambda^2} \right). \quad (3.87)$$

As we are not interested in scalar mesons we will also restrict the field ϕ to the chiral circle $\phi = \langle \phi \rangle = m$. The potential $\mathcal{V}_F[\phi] = \mathcal{V}_F[\langle \phi \rangle]$ is then an irrelevant constant. The real part of the action is then given by $\mathcal{A}_F^R$ (3.84) and the mass term, $\mathcal{A}_m$, see eq. (3.79). Defining $\mathcal{L}_R$ by $\mathcal{A}_m + \mathcal{A}_F^R =: \int d^4x \mathcal{L}_R$ the Lagrangian $\mathcal{L}_R$ is then given by [RD89]

$$\mathcal{L}_R = \frac{1}{2g_V^2} \text{tr} \left((F_V)^2 + (F_A)^2 \right) - 6\frac{m^2}{g_V^2} \text{tr} A^2$$
$$- \frac{1}{4G_2} \text{tr} \left((V - v)^2 + (A - a)^2 \right) + \frac{m_0^u m}{4G_1} \text{tr}(U + U^\dagger - 2) \quad (3.88)$$

where the first line comes from $\mathcal{A}_q^R$ and the second line is the mass term. Remember that $v_\mu = -a_\mu = \frac{1}{2}L_\mu$. So, if we restrict ourselves to $U = 1$ we have $v_\mu = a_\mu = 0$. In order to obtain the standard kinetic energy term of vector and axial–vector fields we renormalize them

$$V_\mu = -ig_V \tilde{V}_\mu; \quad A_\mu = -ig_V \tilde{A}_\mu . \quad (3.89)$$

$\tilde{V}_\mu$ and $\tilde{A}_\mu$ are then the physical vector and axial-vector fields, respectively. Defining

$$\tilde{F}_{\mu\nu}^V = \partial_\mu \tilde{V}_\nu - \partial_\nu \tilde{V}_\mu - ig_V[\tilde{V}_\mu, \tilde{V}_\nu] - ig_V[\tilde{A}_\mu, \tilde{A}_\nu]$$
$$\tilde{F}_{\mu\nu}^A = \partial_\mu \tilde{A}_\nu - \partial_\nu \tilde{A}_\mu - ig_V[\tilde{V}_\mu, \tilde{A}_\nu] - ig_V[\tilde{A}_\mu, \tilde{V}_\nu] \quad (3.90)$$

gives the desired standard form

$$\frac{1}{2g_V^2} \text{tr}(F_V)^2 = -\frac{1}{2} \text{tr}(\tilde{F}_V^2). \quad (3.91)$$

Expanding the vector and axial–vector field in isospin matrices $\tau^a = \{\tau^0 = \mathbb{I}, \tau\}$

$$\tilde{V}_\mu = \tilde{V}_\mu^a \frac{\tau^a}{2} = \omega_\mu \frac{\tau^0}{2} + \rho_\mu \frac{\tau}{2}$$
$$\tilde{A}_\mu = \tilde{A}_\mu^a \frac{\tau^a}{2} = A_\mu^0 \frac{\tau^0}{2} + A_\mu \frac{\tau}{2} \quad (3.92)$$

the Lagrangian (3.88) is for $U = \mathbb{I}$ given by

$$\mathcal{L}_R = -\frac{1}{4}[(\tilde{F}_\nu^a)^2 + (\tilde{F}_A^a)^2] + \frac{1}{2} 6m^2 (\tilde{A}^a)^2$$
$$+ \frac{1}{2}(\frac{g_V^2}{4G_2})[(\tilde{V}^a)^2 + (\tilde{A}^a)^2] . \quad (3.93)$$

The vector and axialvector meson masses are therefore

$$M_V^2 = \frac{g_V^2}{4G_2}$$
$$M_A^2 = M_V^2 + 6M_u^2 . \quad (3.94)$$

Experimentally, we have $M_\rho = 770\text{MeV}$ and $M_\omega = 783\text{MeV}$, *i.e.* the mesons ρ and ω are nearly degenerate which justifies to work in the isospin limit as long as one is not interested in mass differences of the order of a few MeV, and $M_{a_1} \approx 1260\text{MeV}$. As m is the constituent quark mass we expect $m \approx 300\text{MeV}$ – 400MeV which indeed gives approximately the relation (3.94).

It is more transparent to express the parameters in the effective Lagrangian $\mathcal{L}_R$ (3.88) in terms of physical quantities. We therefore define

$$6m^2 = M_A^2 - M_V^2 = M_V^2 \frac{1}{b-1}, \quad b = (1 - \frac{M_V^2}{M_A^2})^{-1} \qquad (3.95)$$

and obtain the effective Lagrangian

$$\mathcal{L}_R = \frac{1}{g_V^2} \text{tr}\left(\frac{1}{2}(F_V^2 + F_A^2) - \frac{M_V^2}{b-1}A^2 + M_V^2\left((V-v)^2 + (A-a)^2\right)\right) \qquad (3.96)$$

$$+ \frac{1}{4}m_\pi^2 f_\pi^2 \text{tr}(U + U^\dagger - 2).$$

3.6 Recovering the Skyrme Model

From the Lagrangian (3.96) we will extract the low–energy pion physics by treating it as effective Lagrangian of the chiral field $\Theta = 2\tau \cdot \pi$. The pion mass m_π is proportional to the current mass m_0, see eq. (3.65), and vanishes in the chiral limit. Therefore we will exploit that $m_\pi^2 \ll M_V^2, M_A^2$. Experimentally the ratio m_π^2/M_V^2 is smaller than $1/30$. Thus, at low energies the pions see the vector and axial–vector mesons as infinitely heavy, *i.e.* static, sources and will not feed back on them. Taking this static limit $M_V^2 \to \infty$ the kinetic energies $(F_V)^2$ and $(F_A)^2$ can be treated as perturbation. Then in lowest order stationary phase approximation the path integral over V_μ and A_μ yields the constraints

$$V_\mu = v_\mu$$
$$A_\mu = \frac{b-1}{b}a_\mu \qquad (3.97)$$

because these conditions extremize the mass terms. Using now $v_\mu = -a_\mu = \frac{1}{2}U^\dagger \partial_\mu U$ and substituting these 'classical' values of V_μ and A_μ in the Lagrangian $\mathcal{L}_R$ we obtain [RD89]

$$\mathcal{L} = -\frac{1}{4}f_\pi^2 \text{tr}(L_\mu L^\mu) + \frac{1}{4}m_\pi^2 f_\pi^2 \text{tr}(U^\dagger + U - 2) + \frac{1}{32e^2}\text{tr}[L_\mu, L_\nu][L^\mu, L^\nu] \qquad (3.98)$$

where

$$f_\pi^2 = \frac{1}{b}\frac{M_V^2}{g_V^2} \qquad (3.99)$$

gives the pion decay constant f_π and

$$m_\pi^2 = \frac{m_0 m}{G_1 f_\pi^2} \qquad (3.100)$$

the pion mass m_π. One sees immediately that eq. (3.100) is equivalent to eq. (3.65), *i.e.* the expression for the pion mass is the same with or without vector mesons. However, the pion decay constant as given by eq. (3.99) is different from the expression given in eq. (3.64). Using eqs. (3.86) and (3.95) it is obvious that both of these expressions for f_π^2 differ by a factor $M_V^2/M_A^2 = 1 - 1/b$. This is due to an effect called $\pi - a_1$–mixing. Due to the chiral left rotation (3.70) we altered implicitly the definition of the chiral field. Without such a rotation we would have obtained terms proportional $\partial_\mu \pi A^\mu$ in the Lagrangian. This would have necessitated a redefinition of the axial–vector and the pion field in order to find the physical modes. Finally, we would have obtained the same result. Simplifying the quark determinant by a chiral rotation we circumvented these problems.

The Lagrangian (3.98) contains also a term which is of fourth order in derivatives. The corresponding parameter

$$e = g_V^2 \frac{b^2}{|2b - 1|} \qquad (3.101)$$

is called Skyrme parameter. The reader certainly has noticed that the Lagrangian (3.98) is exactly the Skyrme model [Sk61]. It is quite remarkable that it can be deduced from NJL–Lagrangian (3.1) using the heat kernel expansion and going to the limit of infinitely heavy vector and axial-vector meson masses.

Let us also remark that for $b = 2$ one finds the Weinberg relation

$$M_A = \sqrt{2} M_V \qquad (3.102)$$

as well as the KSFR relation[§]

$$M_V = \sqrt{2} g_V f_\pi . \qquad (3.103)$$

Using the experimental value of g_V, as *e.g.* determined from the decay $\rho \to 2\pi$, $(g_V)_{\exp} \approx 6.0$, we obtain $e = 4g_V/3 \approx 8$ which has to be compared to the phenomenological value $e \simeq 5$ [Ho93].

The Lagrangian (3.98) contains the free pion Lagrangian which can be seen as follows. First, we expand around the vacuum configuration $\Theta = 0$ for small amplitudes Θ

$$\mathrm{tr}(U^\dagger + U - 2) = 2\mathrm{tr}(\cos\Theta - 1) \simeq 2(-\frac{1}{2}\Theta^2),$$

$$L_\mu = U^\dagger \partial_\mu U \simeq (1 - i\Theta)\partial_\mu(1 + i\Theta) \simeq i\partial_\mu\Theta . \qquad (3.104)$$

[§] For a derivation of these relations see *e.g.* [CL88].

The first two terms of (3.98) read then

$$\mathcal{L}_R = \frac{1}{4} f_\pi^2 \mathrm{tr}(\partial_\mu \Theta \partial^\mu \Theta) - \frac{1}{4} m_\pi^2 f_\pi^2 \mathrm{tr}\Theta^2 + \dots . \qquad (3.105)$$

Defining the physical pion field π as

$$\pi = \tau \cdot \pi = \frac{f_\pi}{2} \Theta \qquad (3.106)$$

we obtain the Lagrangian of an isotriplet of free Klein-Gordon fields, the pions

$$\mathcal{L}_R = \frac{1}{2} \partial_\mu \pi \partial^\mu \pi - \frac{1}{2} m_\pi^2 \pi \pi + \dots . \qquad (3.107)$$

However, the full Skyrme Lagrangian contains also non–linear terms with a larger number of derivatives. It describes pions for small energies and amplitudes reasonably well. The self–interaction of the pion field contained in this Lagrangian allows for special localized solutions: topological solitons. Skyrme introduced this model already 1960 out of purely phenomenological reasons without knowing anything about quarks or QCD [Sk61]. In this model baryons are described as solitons of meson fields. In Chap. 4 we will discuss the description of baryons as solitons within QFD, see there for more details. In the next section we will discuss first the imaginary part of the quark loop.

3.7 Chiral Anomaly

3.7.1 Non–invariance of the Path Integral Measure

In a quantum theory there are three distinct possibilities of breaking a symmetry. First, there is explicit breaking by adding a non–invariant term to the (classical) Lagrangian. Non–vanishing current masses are an example for explicit breaking of chiral symmetry. Second, there is spontaneous breakdown of a symmetry which means that the Lagrangian is symmetric, however, the states, especially the ground state, are not invariant under the related symmetry transformation. An example easy to visualize is the ferromagnet. Also the breaking of chiral symmetry discussed in sect. 3.3 is a typical example for this phenomenon. Third, there is anomalous breaking of a symmetry. This means that a symmetry present in the classical limit $\hbar \to 0$ is explicitly violated in the quantum theory. The classical Lagrangian is invariant, however, the symmetry is spoiled by quantum fluctuations. Defining quantum mechanics via path integrals anomalous symmetry breaking is realized due to the non–invariance of the functional integral measure [Fu80] despite the fact that the classical action appearing as a 'weight' factor under the path integral is invariant under the considered symmetry. We will discuss this for anomalous chiral symmetry breaking following the work of Fujikawa [Fu80].

The path integral formulation of anomalous chiral symmetry breaking can be summarized as follows: The integral measure

$$\mathcal{D}q\mathcal{D}\bar{q} \qquad (3.108)$$

is not invariant under chiral rotations

$$q(x) \rightarrow \tilde{q}(x) = U(x)q(x) = e^{i\Theta(x)\gamma_5}q(x), \quad \Theta = \Theta^a \left(\frac{\lambda^a}{2}\right)_F$$

$$\bar{q}(x) \rightarrow \bar{\tilde{q}}(x) = \bar{q}(x)U(x) \tag{3.109}$$

but acquires a phase $J(\Theta) \neq 1$

$$\mathcal{D}\tilde{q}\mathcal{D}\bar{\tilde{q}} = J(\Theta)\mathcal{D}q\mathcal{D}\bar{q}. \tag{3.110}$$

In order to make the proof of anomalous chiral symmetry breaking, reflected by the relation $J(\Theta) \neq 1$, more transparent we will use in the following a few simplifications. First, we will only consider infinitesimal chiral transformations, i.e. small Θ. For simplicity we consider only the abelian subgroup $U_A(1)$. Therefore we assume from the beginning that Θ is proportional to the unit matrix in flavor space, $\Theta \propto \mathbb{I}_F$. This assumption will make the formulas more transparent as Θ commutes with other flavor matrices. The whole calculation may be done keeping these commutators with the result that the traceless parts of Θ have to vanish. Second, we will work with a Dirac operator containing besides the usual kinetic and mass term only a vector field

$$i\slashed{D} = i(\slashed{\partial} + \slashed{V}) - m_0 =: i\slashed{d} - m_0. \tag{3.111}$$

As we are working with antihermitian Dirac matrices and vector fields the operator $i\slashed{d}$ is hermitian with respect to the ordinary scalar product

$$\langle\varphi_1|\varphi_2\rangle = \int d^4x\varphi_1^\dagger(x)\varphi_2(x). \tag{3.112}$$

Therefore the eigenfunctions of the operator $i\slashed{d}$ form a complete and orthogonal set which can be chosen normalized

$$i\slashed{d}\varphi_n(x) = \lambda_n\varphi_n(x)$$

$$\int d^4x\varphi_n^\dagger(x)\varphi_m(x) = \delta_{nm}$$

$$\sum_n \varphi_n(x)\varphi_n^\dagger(y) = \delta(x-y). \tag{3.113}$$

We expand the dynamical quark field, i.e. the functional integration variable, in terms of the functions φ_n

$$q(x) = \sum_n a_n\varphi_n(x)$$

$$\bar{q}(x) = \sum_n a_n^*\bar{\varphi}_n(x) \tag{3.114}$$

The expansion coefficients a_n and a_n^* are anticommuting Grassmann variables. They are connected via an operation called involution $a_n \rightarrow a_n^*$ and $a_n^* \rightarrow a_n$.

Note that this operation is different from complex conjugation. The integration measure can now be written as

$$\mathcal{D}\bar{q}\mathcal{D}q = \prod_m da_m da_m^* \tag{3.115}$$

where the Grassmann variables a_m and a_m^* are independent integration variables similar to the integration over complex numbers where z and z^* or, alternatively, real and imaginary part are independent variables with respect to the integration measure. The chirally transformed quark field is given by

$$\tilde{q}(x) = \sum_m \tilde{a}_m \varphi_m(x) = \sum_m a_m e^{i\Theta(x)\gamma_5} \varphi_m(x)$$

$$\tilde{\bar{q}}(x) = \sum_m \bar{\varphi}_m(x)\tilde{a}_m^* = \sum_m \bar{\varphi}_m(x) a_m^* e^{i\Theta(x)\gamma_5}. \tag{3.116}$$

where the expansion coefficients for the transformed quark field are related to the original ones by

$$\tilde{a}_m = \sum_n C_{mn} a_n$$

$$\tilde{a}_m^* = \sum_n a_n^* C_{nm}$$

$$C_{mn} = \langle m|e^{i\Theta(x)\gamma_5}|n\rangle = \int d^4x\, \varphi_m^\dagger(x) e^{i\Theta(x)\gamma_5} \varphi_n(x)$$

$$= \delta_{mn} + i\int d^4x\, \Theta(x) \varphi_m^\dagger(x)\gamma_5 \varphi_n(x) + \mathcal{O}(\Theta^2). \tag{3.117}$$

Here we have used the orthonormality of the eigenfunctions φ_n. As a_n, a_n^* and $\tilde{a}_m, \tilde{a}_m^*$ are Grassmann variables we obtain

$$J(\Theta) = (\mathrm{Det}C)^{-2}. \tag{3.118}$$

For infinitesimally small Θ the matrix C is given by

$$C_{mn} = \delta_{mn} + \langle m|i\gamma_5\Theta(x)|n\rangle + \mathcal{O}(\Theta^2). \tag{3.119}$$

Defining now

$$\epsilon_{mn} = \langle m|i\gamma_5\Theta(x)|n\rangle \tag{3.120}$$

and writing shorthand $C = 1 + \epsilon$ the functional determinant of this matrix is evaluated to be in first order in Θ

$$\begin{aligned}
\mathrm{Det}C &= \mathrm{Det}(1+\epsilon) \\
&= \exp\left(\mathrm{Tr}\log(1+\epsilon)\right) \\
&= \exp\left(\mathrm{Tr}(\epsilon) + \mathcal{O}(\epsilon^2)\right) \\
&= \exp\left(i\int d^4x\,\Theta(x)\sum_n \varphi_n^\dagger \gamma_5 \varphi_n(x)\right) + \dots.
\end{aligned} \tag{3.121}$$

The operator appearing in the exponent is not well–defined as it stands

$$B(x) := \sum_n \varphi_n^\dagger \gamma_5 \varphi_n(x) = \text{tr}(\gamma_5) \sum_n \varphi_n \varphi_n^\dagger(x) = \text{tr}(\gamma_5)\delta(0) = 0 \times \infty$$

$$(3.122)$$

where we have used the completeness of the functions φ_n and the tracelessness of γ_5. In order to make $B(x)$ well–defined we have to complete the calculation within some regularization and to remove the regularization at the end of the calculation

$$\begin{aligned}
B(x) &:= \lim_{\Lambda\to\infty} \sum_n \varphi_n^\dagger \gamma_5 e^{-(\lambda_n/\Lambda)^2} \varphi_n(x) \\
&= \lim_{\Lambda\to\infty} \sum_n \varphi_n^\dagger \gamma_5 e^{-\slashed{A}^\dagger \slashed{A}/\Lambda^2} \varphi_n(x) \\
&= \lim_{\Lambda\to\infty} \lim_{x\to y} \text{tr}\, \gamma_5 \, \langle x|e^{-\slashed{A}^\dagger \slashed{A}/\Lambda^2}|y\rangle \\
&= \lim_{\Lambda\to\infty} \text{tr}\, \gamma_5 \, \langle x|e^{-\slashed{A}^\dagger \slashed{A}/\Lambda^2}|x\rangle.
\end{aligned} \tag{3.123}$$

In order to apply the heat kernel expansion of Appendix A we introduce the "proper time" $\tau = 1/\Lambda^2$ and use as unperturbed operator $A_0 = \partial_\mu \partial^\mu + \mu^2$ (A.9). Expanding now the relevant matrix element

$$\langle x|e^{-\tau \slashed{A}^\dagger \slashed{A}}|y\rangle = \langle x|e^{-\tau A_0}|y\rangle \sum_{k=0}^\infty h_k(x,y)\tau^k \tag{3.124}$$

and using eq. (A.10)

$$\langle x|e^{-\tau A_0}|y\rangle = \frac{1}{(4\pi\tau)^2} e^{(x-y)^2/4\tau - \mu^2\tau} \tag{3.125}$$

we obtain

$$B(x) = \frac{1}{(4\pi)^2} \lim_{\tau\to 0} \sum_{k=0}^\infty \tau^{k-2} \text{tr}(\gamma_5 h_k(x,x)). \tag{3.126}$$

As the operator considered here is even simpler than the operator A used in Appendix A we may use the heat coefficients h_k (A.22), (A.28) and (A.29) with axialvector and scalar field set to zero, $A_\mu = \phi = 0$. Note that only the heat coefficients h_0, h_1 and h_2 in the sum (3.126) do contribute to $B(x)$ because the coefficients are suppressed by a factor τ^{k-2} and therefore the higher order terms vanish in the limit $\tau \to 0$ ($\Lambda \to \infty$). As

$$\text{tr}\, \gamma_5\, h_0(x,x) = \text{tr}\, \gamma_5 = 0$$

$$\text{tr}\, \gamma_5\, h_1(x,x) = -\frac{1}{2}\text{tr}(\gamma_5 \gamma_\mu \gamma_\nu)\Gamma_{\mu\nu} = 0 \tag{3.127}$$

only one term contributes

$$\text{tr}\gamma_5 h_2(x,x) = \frac{1}{2}\text{tr}_c\text{tr}_D\text{tr}_F\gamma_5 a^2$$

$$= \frac{N_c}{8}\text{tr}_D(\gamma_5\gamma^\mu\gamma^\nu\gamma^\kappa\gamma^\lambda)\text{tr}_F\Gamma_{\mu\nu}\Gamma_{\kappa\lambda}$$

$$= \frac{N_c}{2}\epsilon^{\mu\nu\kappa\lambda}\text{tr}_F\Gamma_{\mu\nu}\Gamma_{\kappa\lambda}$$

$$= N_c\text{tr}_F\tilde{\Gamma}^{\mu\nu}\Gamma_{\mu\nu} \qquad (3.128)$$

where we used the definition of the dual tensor

$$\tilde{\Gamma}^{\mu\nu} = \frac{1}{2}\epsilon^{\mu\nu\kappa\lambda}\Gamma_{\kappa\lambda}. \qquad (3.129)$$

Eqs. (3.126) and (3.128) can now be combined to the result

$$B(x) = \frac{N_c}{16\pi^2}\text{tr}_F\tilde{\Gamma}_{\mu\nu}\Gamma^{\mu\nu} \qquad (3.130)$$

finally yielding the desired expression for the Jacobian $J(\Theta)$ (see eqs. (3.117) and (3.118))

$$J(\Theta) = (\text{Det}C)^{-2} = \exp\left(-i\frac{N_c}{8\pi^2}\int d^4x\Theta(x)\text{tr}_F(\tilde{\Gamma}\Gamma)\right). \qquad (3.131)$$

Note that the Jacobian $J(\Theta)$ is a pure phase factor, $|J(\Theta)| = 1$, and therefore contributes to the imaginary part of the effective action. In the presence of a vector field this factor will be in general not equal to unity which concludes our proof of the axial anomaly.

3.7.2 Some Remarks on the Chiral Anomaly

Using the invariance of the classical action we can easily derive the so-called anomalous Ward identity. Under infinitesimally small chiral rotations the quark bilinear in the action transforms as

$$\bar{q}(i\slashed{\partial} - m_0)q = \bar{\tilde{q}}e^{-i\Theta(x)\gamma_5}(i\slashed{\partial} - m_0)e^{-i\Theta(x)\gamma_5}\tilde{q}$$

$$= \bar{\tilde{q}}(i\slashed{\partial} - m_0)\tilde{q}$$

$$+ \partial_\mu\Theta(x)\bar{\tilde{q}}\gamma^\mu\gamma_5\tilde{q} + 2im_0\Theta(x)\bar{\tilde{q}}\gamma_5\tilde{q} + \mathcal{O}(\Theta^2) \qquad (3.132)$$

which leads to the following expressions for the effective quark action

$$\log Z_q = \text{Tr}\log(i\slashed{\partial} - m_0)$$

$$= \log\int \mathcal{D}q\mathcal{D}\bar{q}\exp\left(i\int d^4x\bar{q}(i\slashed{\partial} - m_0)q\right)$$

$$= \log J(\Theta) \qquad (3.133)$$

$$+ \log\int \mathcal{D}\tilde{q}\mathcal{D}\bar{\tilde{q}}\exp\left(i\int d^4x\bar{\tilde{q}}(i\slashed{\partial} - m_0)\tilde{q}\right.$$

$$\left. + \partial_\mu\Theta(x)\bar{\tilde{q}}\gamma^\mu\gamma_5\tilde{q} + 2im_0\Theta(x)\bar{\tilde{q}}\gamma_5\tilde{q}\right).$$

As the l.h.s. is independent of Θ the r.h.s. has to be, too. Therefore one obtains

$$0 = \frac{\delta \log Z_q}{\delta \Theta}\Big|_{\Theta=0} = -i\frac{N_c}{8\pi^2}\langle \operatorname{tr}_F(\tilde{\Gamma}\Gamma)\rangle - \partial_\mu\langle \bar{\tilde{q}}\gamma^\mu\gamma_5\tilde{q}\rangle + 2im_0\langle\bar{\tilde{q}}\gamma_5\tilde{q}\rangle. \quad (3.134)$$

Defining now the currents

$$j_5^\mu = \langle \bar{\tilde{q}}\gamma^\mu\gamma_5\tilde{q}\rangle$$
$$j_5 = \langle \bar{\tilde{q}}\gamma_5\tilde{q}\rangle \quad (3.135)$$

we arrive at the anomalous Ward identity

$$\partial_\mu j_5^\mu = 2im_0 j_5 - i\frac{N_c}{8\pi^2}\operatorname{tr}_F\langle(\tilde{\Gamma}\Gamma)\rangle. \quad (3.136)$$

Note that even for $m_0 = 0$ the axial current j_5^μ is not conserved if vector fields are present despite the fact that the classical Lagrangian is invariant under chiral rotations for $m_0 = 0$. This reflects the anomalous symmetry breaking, *i.e.* symmetry breaking by quantum effects (here of the quarks).

Noether's theorem states that the corresponding current j_5^μ should be conserved for $m_0 = 0$. The violation of Noether's theorem indiates therefore anomalous symmetry breaking. Note also that the term $\tilde{\Gamma}\Gamma$ can be written as the divergence of a current and should only give a contribution at the 'surface' of space–time. However, the corresponding current is gauge dependent which leads to observable effects of such a term. For a pedagogical discussion the reader is refered to [Co90].

The axial anomaly as formulated by the anomalous Ward identity (3.136) is known as Adler–Bell–Jackiw anomaly [ABJ69]. Usually it is required that in fundamental (gauge) theories anomalies should be not present or should be canceled by other effects. As we are considering effective low–energy models anomalies can be present. They even have measurable consequences as *e.g.* the decay $\pi^0 \to 2\gamma$.

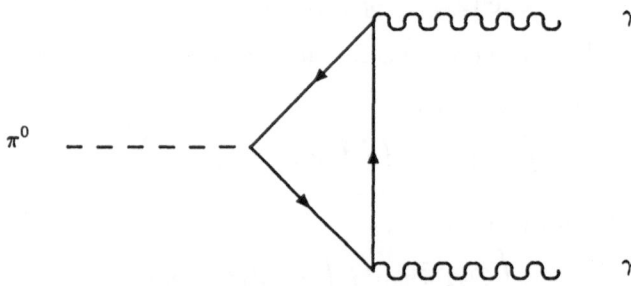

Fig. 3.7. The triangle diagram which is responsible for the anomaly.

Historically, the anomaly was discovered in perturbation theory. The one–loop diagram which is responsible for the decay $\pi^0 \to 2\gamma$ is shown in Fig. 3.7. The path integral derivation of the anomaly is a non–perturbative one and demonstrates that higher order terms do **not** contribute to the anomaly. Due to some misunderstandings in the literature [BHS88] we want to phrase this more precisely: If one calculates the Adler–Bell–Jackiw anomaly in a renormalizable theory (as QCD is) in which the cutoff Λ is taken to infinity at the end of the calculation only the triangle diagram Fig. 3.7 contributes. However, if one works with dressed quarks, *i.e.* with finite constituent quark masses, within an effective non–renormalizable theory where the cutoff has to be kept finite not only the triangle but all higher order diagrams contribute. Remember that the limit $\Lambda \to \infty$, *i.e.* $\tau \to 0$, was necessary in order to make the contributions from h_n, $n \geq 3$ vanish. In the literature there are claims that the NJL model is not appropiate for the treatment of the anomaly and the π^0 decay because the triangle diagrams gives only about two thirds of the experimentally determined decay amplitude using a finite cutoff [BHS88]. However, together with the finite cutoff also higher order terms should be included and yield the missing third of the decay amplitude.

3.7.3 The Wess–Zumino–Witten Action

Until now we have only calculated the Jacobian for infinitesimal Abelian chiral rotations

$$\frac{\delta J(\Theta)}{\delta \Theta} = -i \frac{N_c}{8\pi^2} \mathrm{tr}_F (\tilde{\Gamma}_{\mu\nu} \Gamma^{\mu\nu}). \tag{3.137}$$

Additionally, we want to know the Jacobian $J(\Theta)$ for finite non–abelian chiral transformations

$$U(x) = e^{i\Theta(x)}, \quad \Theta(x) = \Theta^a(x)\frac{\lambda^a}{2}. \tag{3.138}$$

We are going to demonstrate that it can be calculated by functional integration of the anomaly

$$J(\Theta) = \int \mathcal{D}\Theta \frac{\delta J(\Theta)}{\delta \Theta}$$
$$= \exp(i\mathcal{A}_{WZW}). \tag{3.139}$$

This term $\mathcal{A}_{WZW}$ being the logarithm of the integrated anomaly is the famous Wess–Zumino–Witten (WZW) term [Wi79], the field theoretical analogue of the Berry phase [Be84]: As $\mathcal{A}_{WZW}$ is real the integrated anomaly gives a pure phase factor similar to the quantum–mechanical Berry phase. As we will see it is closely connected to the topological nature of the chiral field $\Theta(x)$. In general it is a very complicated functional. We will treat therefore for pedagogical transparency only the special case of a flavor singlet vector meson (ω–meson) and a SU(2) chiral field

$$V_\mu(x) = \omega_\mu(x)\,\mathbb{I}_F$$
$$U(x) = e^{i\Theta(x)}, \quad \Theta(x) = \Theta^a(x)\frac{\tau^a}{2}. \tag{3.140}$$

The WZW term is then given by

$$\mathcal{A}_{WZW} = N_c \int d^4x\, \omega_\mu(x) B^\mu(x) \tag{3.141}$$

where

$$B^\mu(x) = \frac{1}{24\pi^2}\epsilon^{\mu\nu\kappa\lambda}\mathrm{tr}(L_\nu L_\kappa L_\lambda), \quad L_\nu = U^\dagger \partial_\nu U \tag{3.142}$$

is a topological current. It is conserved due to entirely algebraic properties (antisymmetry of the Levi–Cevita symbol). Since $U^\dagger U = 1$ we have

$$(\partial_\mu U^\dagger)U + U^\dagger \partial_\mu U = 0. \tag{3.143}$$

Using this and the antisymmetry of the ϵ–symbol one can easily show that

$$\partial_\mu B^\mu = -\frac{1}{8\pi^2}\epsilon^{\mu\nu\kappa\lambda}\mathrm{tr}(L_\mu L_\nu L_\kappa L_\lambda). \tag{3.144}$$

The cyclic property of the trace implies then that the above expression is equal to its negative and therefore

$$\partial_\mu B^\mu = 0 \tag{3.145}$$

i.e. the current B^μ is conserved independent of the explicit form of the chiral field $\Theta(x)$.

In order to reveal the physical nature of this current we consider the quark generating functional treating the vector field V_μ as an external source. Performing a chiral rotation and exploiting the relation (3.110) one obtains

$$
\begin{aligned}
Z_q &= \int \mathcal{D}q\mathcal{D}\bar{q}\exp\left(\int d^4x\,\bar{q}(i(\partial\!\!\!/ + V\!\!\!\!/) - m^0)q\right) \\
&= J(\Theta)e^{\mathcal{A}[V]} \\
&= \exp\left(iN_c\int d^4x\, V_\mu(x)B^\mu(x) + \mathcal{A}[V]\right)
\end{aligned}
\tag{3.146}
$$

where $\mathcal{A}[V]$ contains the non–anomalous terms of the effective action. From the first line we obtain

$$\frac{1}{N_c}\left(\frac{\delta \log Z_q}{i\delta V_\mu(x)}\right)_{V=0} = \frac{1}{N_c}\langle\bar{q}(x)\gamma^\mu q(x)\rangle =: j_B^\mu(x), \tag{3.147}$$

i.e. the baryon current of the quarks. On the other hand, the last line gives as contribution from the anomaly the current $B^\mu(x)$, *i.e.*

$$j_B^\mu(x) = B^\mu(x) + \ldots \tag{3.148}$$

where the dots indicate the contributions from the "normal" terms $\mathcal{A}[V]$. This leads to the following interpretation: The baryon current of the original fermion (quark) theory is carried by the topological current of the chiral field in the bosonized effective meson theory, at least in a leading order gradient expansion. As we have seen this topological current is conserved independent of the specific form of the chiral field whereas the baryon current of the underlying quark theory is the (conserved) Noether current of flavor singlet phase transformations. So, the dynamical property of the quark theory has turned into a purely topological property in the effective theory. We summarized some basic topological features of the chiral field in Appendix B.

These considerations lead immediately to the question how the pseudoscalar meson field $\Theta(x)$ which is bosonic and spinless can describe properties of baryons which are fermions with spin $s = 1/2$. We will demonstrate in the next chapter that a sufficiently strong topological non–trivial chiral field with **winding number** n (see Appendix B) polarizes the vacuum or Dirac sea of the quarks so strongly that n of the valence quark levels are so tightly bound that they join the negative Dirac sea: their energy is negative. As the physical vacuum is defined as the state with lowest energy and therefore in the vacuum all negative energy levels are occupied the physical vacuum then carries a non–vanishing baryon number

$$N_B := \int d^3x \, B^0(x) = n. \tag{3.149}$$

So it is not really the chiral field itself which carries the baryon number but the polarized vacuum. As we cannot observe the vacuum but only the polarizing meson field we relate the baryon number to the chiral field. In that sense meson fields carry baryonic charge. This fact is the underlying feature for the description of the baryons as chiral solitons which we will discuss in detail in the next chapter.

Summary

The quark theory based on the Lagrangian (3.1) is transformed with the help of functional integral techniques to a purely mesonic theory. Expanding the corresponding action in the fluctuations of the meson fields around their vacuum expectation values to first and second order yields the formal expressions for the Dyson–Schwinger and Bethe–Salpeter equations, respectively. The solution of the Dyson–Schwinger equation in the scalar and pseudoscalar channel reveals that chiral symmetry is broken dynamically within QFD. This becomes evident also when considering the effective potential: The chiral symmetry breaking vacuum is energetically favored as compared to the pertubative vacuum. An explicit expansion of the pion field up to second order provides via the Bethe–Salpeter equation an expression for the pion mass and decay constant. These two quantities are related to the quark condensate and current mass such that the Gell-Mann–Renner–Oakes relation (3.65) is fulfilled to first order in the current mass.

When including vector and axialvector mesons it is convenient to use a chiral rotation to remove the pseudoscalar field from the mass term in the quark determinant. Using additionally a gradient expansion one arrives at the gauged linear σ–model (3.96). On the chiral circle and in the limit of infinitely heavy vector and axialvector mesons this model reduces to the Skyrme model with the Skyrme parameter e given in terms of (axial) vector meson properties.

The functional integral measure is not invariant under flavor singlet chiral rotation, *i.e.* this symmetry is broken in an anomalous fashion. Calculating the variation of the measure under this chiral rotation gives the Jacobian, which in turn provides the imaginary part of the action, as well as the anomalous Ward identity (3.136). Finally, the Wess–Zumino–Witten action is obtained by integrating the infinitesimal anomaly. Its dependence on the topological current has lead us to the following conjecture: The baryon current of the original quark theory is carried by the topological current in the bosonized effective meson theory.

Chapter 4

Baryons as Chiral Solitons

4.1 Basic Aspects of the Skyrmion

In Sect. 3.5 we derived the Skyrme Lagrangian (3.98)

$$\mathcal{L} = -\frac{1}{4}f_\pi^2 \text{tr}(L_\mu L^\mu) + \frac{1}{4}m_\pi^2 f_\pi^2 \text{tr}(U^\dagger + U - 2) + \frac{1}{32e^2}\text{tr}[L_\mu, L_\nu][L^\mu, L^\nu]$$

$$U = e^{2i\pi}, \quad \pi = \pi^i \frac{\tau^i}{2}, \quad L_\mu = U^\dagger \partial_\mu U \tag{4.1}$$

from the non–local effective meson theory (3.26) which is equivalent to the Nambu–Jona-Lasinio model (3.1). We achieved this by using a covariant gradient expansion, the heat kernel method, and taking the limit of infinitely large vector and axialvector meson masses. We discussed already that the Skyrme model describes pions for small energies and amplitudes reasonably well. Allowing also large amplitudes one realizes that this model possesses special localized solutions: topological solitons. These are then interpreted as baryons.

The idea to describe baryons as solitons was first formulated by Skyrme more than thirty years ago [Sk61]. This work was not fully appreciated until Witten conjectured that baryons arise as solitons of the meson fields in large N_c QCD [Wi79]. Since then a lot of papers appeared discussing various aspects of the Skyrmion (for a review see *e.g.* [Ho93]). Here we will restrict ourselves to a few basic concepts.

In order to find a solitonic solution it is sufficient to consider static fields. The chiral field, *i.e.* the flavor matrix, U describes then a map from ordinary space into the flavor group $SU(N_f)$. A fundamental requirement for a solitonic solution of the equations of motion is finiteness of energy. By inspection of the Lagrangian (4.1) one realizes that a necessary and sufficient condition to achieve this is to use a chiral field U that approaches a constant at spatial infinity. For convenience we require

$$\lim_{r \to \infty} U(x) = 1. \tag{4.2}$$

This means that spatial infinity is mapped to one point in flavor space implying that all points at spatial infinity are identical, *i.e.* space $I\!R^3$ is compactified to the hyperspere S^3. In Appendix B it is shown that such fields can be characterized by their winding number, *i.e.* a topological property, and that this winding number is equal to the baryon number as defined in eq. (4.40).

From now on we use the spherically symmetric hedgehog *ansatz*

$$U(\boldsymbol{x}) = \exp\big(i\boldsymbol{\tau} \cdot \hat{\mathbf{r}}\,\Theta(r)\big). \tag{4.3}$$

This configuration has neither good angular momentum nor good isospin. However, it has vanishing 'grand spin' $G = l + \sigma/2 + \tau/2$, *i.e.* $[G, U] = 0$. Note that the grand spin is the sum of the isospin $\tau/2$ and the total angular momentum j which in turn is the sum of spin and orbital angular momentum, $j = l + \sigma/2$. The spatial components of the left current of the hedgehog field are given by

$$\begin{aligned}
L_i &= U^\dagger \nabla_i U \\
&= (\cos\Theta - i\boldsymbol{\tau} \cdot \hat{\mathbf{r}}\sin\Theta) \\
&\quad \left(-\hat{r}_i \Theta' \sin\Theta + i\tau_i \frac{\sin\Theta(r)}{r} + i(\boldsymbol{\tau} \cdot \hat{\mathbf{r}})\hat{r}_i\left(\Theta'\cos\Theta - \frac{\sin\Theta}{r}\right)\right)
\end{aligned} \tag{4.4}$$

where $\Theta' = d\Theta/dr$. Using this in eq. (B.7) for the winding number (which is identified with the baryon number) one obtains after some lengthy, however, straightforward calculation

$$\begin{aligned}
N_B &= \frac{1}{2\pi^2} \int d^3x\,\Theta' \frac{\sin^2\Theta}{r^2} \\
&= \frac{1}{\pi}\Big(\Theta(r) + 2\sin\Theta(r)\cos\Theta(r)\Big)\big|_0^\infty
\end{aligned} \tag{4.5}$$

Choosing $\Theta(0) = -\pi$ and $\Theta(\infty) = 0$ gives $N_B = 1$.

The energy of the static soliton depends on the chiral field $\Theta(r)$ and its derivative Θ'

$$E = 4\pi \int_0^\infty dr \left(\frac{1}{2}f_\pi^2(r^2 + (\Theta')^2 + 2\sin^2\Theta) + \frac{1}{64e^2}\sin^2\Theta(2(\Theta')^2 + \frac{\sin^2\Theta}{r^2})\right). \tag{4.6}$$

Variation with respect to Θ gives the equation of motion which is a second order differential equation

$$\Theta''\left(\tilde{r}^2 + 2\sin^2\Theta\right) + 2\Theta'\tilde{r} + (\Theta'^2 - 1)\sin 2\Theta - \frac{\sin^2\Theta\sin 2\Theta}{\tilde{r}^2} = 0 \tag{4.7}$$

where $\tilde{r}$ denotes the dimensionless variable $\sqrt{32}ef_\pi r$ and the symbol ' has be redefined accordingly

$$\Theta'(\tilde{r}) = \frac{d\Theta(\tilde{r})}{d\tilde{r}}. \tag{4.8}$$

Using the boundary conditions $\Theta(0) = -\pi$ and $\Theta(\infty) = 0$ the equation of motion can be solved numerically.

It is possible not to take the limit of infinitely heavy vector and axialvector meson masses. As a matter of fact a Lagrangian containing explicitely (axial) vector mesons, as *e.g.* the one defined in eq. (3.96) also exhibits solitonic solutions which are also called Skyrmions in the literature, see *e.g.* [MKW87]. In these lecture notes we will not go into more detail as the Skyrmion is concerned. In the following we will concentrate on the question how the underlying quark structure of the baryons is represented in the picture of baryons as chiral solitons.

4.2 Diving into the Dirac Sea

The Skyrmion picture of baryons rests on the identification of the baryon number with the winding number. Whereas in Skyrme type, *i.e.* purely mesonic, models this is the underlying assumption one may test this hypothesis for the effective meson theory (3.26). In order to arrive at the Skyrme model we had to do a covariant gradient expansion. Obviously, this expansion is only justified for small momenta as compared to the cutoff. However, it is possible to evaluate the fermion determinant (3.68) exactly, *i.e.* numerically, for a baryon number one configuration. As the non–local meson theory is exactly equivalent to the Nambu–Jona-Lasinio model (3.1) this solitonic solution is often called the soliton of the Nambu–Jona-Lasinio model or NJL soliton.

In order to analyze the quark spectrum it is convenient to allow only the chiral field U to deviate from its VEV, *i.e.* we "freeze" the other fields: scalar, vector and axial–vector. For the chiral field we choose the hedgehog ansatz (4.3) supplemented by an ansatz for the chiral angle

$$\Theta(r) = -\pi \begin{cases} 1 - \frac{2r}{3a} & \text{for } r \leq a \\ \frac{a^2}{3r^2} & \text{for } r \geq a \end{cases}. \tag{4.9}$$

This special form is motivated by the fact that for small r the chiral angle is linear whereas in the chiral limit it is proportional to $1/r^2$ for large r, see Sect. 4.4. The coefficients and the matching point a are then chosen in a way that the chiral angle and its derivative, $\Theta(r)$ and $\Theta'(r)$, are continuous [MR86].

As we will see in the next section the one–particle Dirac Hamiltonian

$$h = \boldsymbol{\alpha} \cdot \boldsymbol{p} + \beta m (\cos\Theta(r) + i\gamma_5 \boldsymbol{\tau} \cdot \hat{\mathbf{r}} \sin\Theta) \tag{4.10}$$

represents the quark energy spectrum in the presence of the chiral hedgehog field. Diagonalizing (4.10) for the chiral angle (4.9) for different values of the parameter a provides the following picture (see Fig. 4.1): For a very small $a \ll 1/m$, *i.e.* for a small chiral angle almost everywhere, the spectrum is only slightly distorted as compared to the unperturbed one.[*] If a increases the lowest positive energy level starts to decrease and for $a \gg 1/m$ it joins the Dirac sea. As this level

---

[*] In order to obtain a discrete spectrum the system is treated in a finite spherical box with radius $D \gg 1/m$. For numerical details see [KR84, AWZ94].

crosses the zero–energy line it is filled as the vacuum state is defined to be the one with lowest energy. Therefore the distorted Dirac vacuum carries the baryon number.** The baryon number given by

$$B = -\frac{1}{2} \sum_\mu \text{sign}\epsilon_\mu \qquad (4.11)$$

(with ϵ_μ being the eigenvalues of h (4.10)) is then equivalent to the winding number corresponding to the chiral field (4.9).

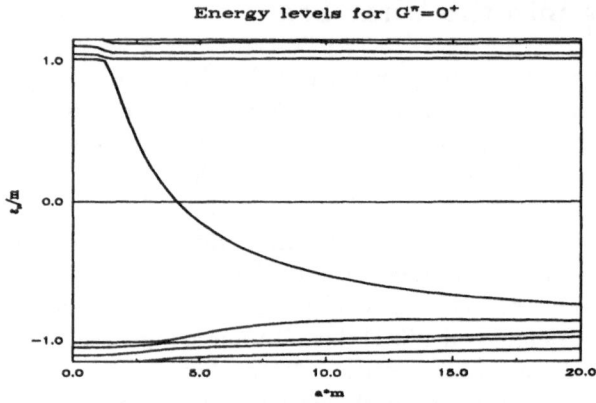

Fig. 4.1. The energy levels for grand spin zero and positive parity for the chiral field (4.9) as function of the parameter a. Energies and lenghts are given in units of the constituent quark mass m.

It is now quite obvious that *e.g.* for

$$\Theta(r) = -2\pi \begin{cases} 1 - \frac{2r}{3a} & \text{for } r \le a \\ \frac{a^2}{3r^2} & \text{for } r \ge a \end{cases} \qquad (4.12)$$

which carries winding number $n = 2$ two levels are pulled down, see Fig. 4.2. The only thing which is probably surprising is that the second "diving" level has negative parity. On the other hand it is the second lowest level (in a finite box).

In summary, we have demonstrated that the identification of the topological winding number with the baryon number, *i.e.* Witten's conjecture [Wi79], arises naturally in Quark Flavor Dynamics if the solitonic meson fields are strong enough to pull positive–energy quark levels into the Dirac sea. That this is the case for the self–consistently determined soliton of Quark Flavour Dynamics will be demonstrated in the next sections.

** One might argue that introducing a chemical potential $m < \mu < 0$, *i.e.* shifting the zero–energy line from the middle of the gap to the lower half of the gap, one might escape this interpretation. Note, however, that for $a \gg 1/m$ this "valence level" becomes indistinguishable from the negative sea.

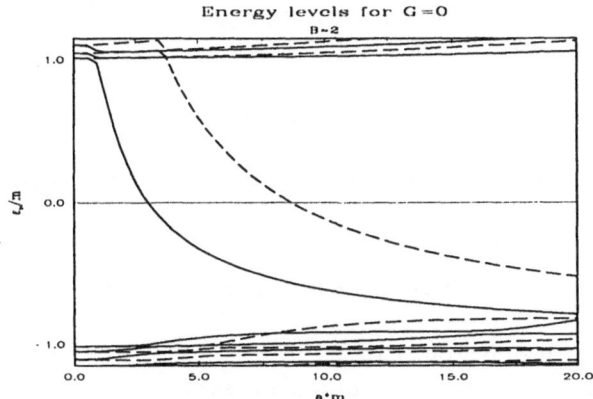

Fig. 4.2. The energy levels for grand spin zero (positive parity levels are shown as full lines, negative parity ones as dashed lines) for the chiral field (4.12) as function of the parameter a. Energies and lenghts are given in units of the constituent quark mass m.

4.3 The Energy Functional of the NJL Soliton for Static Meson Fields

In a first step towards a self–consistent soliton solution of the non–local effective meson theory (3.26) we will derive the energy functional of the static soliton. First, we will consider the fermion determinant (3.68) using the separation into real and imaginary part (3.80). As we are interested in static solutions we write the Dirac operator $\not{D}$ as a sum of a time derivative and a one particle Dirac Hamiltonian, $i.e.$ after Wick rotation the Euclidian Dirac operator is given by

$$i\beta\not{D} = -\partial_\tau - h,$$
$$h = \boldsymbol{\alpha}\cdot\boldsymbol{p} + iV_4 + i\gamma_5 A_4 + \boldsymbol{\alpha}\cdot\boldsymbol{V} + \gamma_5\boldsymbol{\alpha}\cdot\boldsymbol{A} + \beta(P_R M + P_L M^\dagger) \quad (4.13)$$

wherein τ denotes the Euclidian time and for the time component of the vector and axialvector field we have used $V^0 = -iV_4$, $A^0 = -iA_4$. In Euclidian space τ, V_4 and A_4 have to be considered as real quantities. This leads to a non–Hermitean Hamiltonian h even for static configurations ($i.e.$ $[\partial_\tau, h] = 0$) if non–vanishing time components of vector or axialvector meson fields are included.

For the chiral field we impose again the hedgehog ansatz:

$$U(\boldsymbol{x}) = \exp\left(i\boldsymbol{\tau}\cdot\hat{\mathbf{r}}\,\Theta(r)\right). \quad (4.14)$$

As this configuration has vanishing grand spin $\boldsymbol{G} = \boldsymbol{l}+\boldsymbol{\sigma}/2+\boldsymbol{\tau}/2$ the other meson fields should have this property, too. Also we want to keep parity. Then for the scalar field only the isoscalar part may deviate from its vacuum expectation value m and it may depend only on the radius r,

$$\Phi(\boldsymbol{x}) = \phi(r)\mathbb{1}. \quad (4.15)$$

The only possible ansatz for the ω field with grand spin zero has vanishing spatial components($\omega_i = 0$):

$$\omega_\mu = \omega(r)\delta_{\mu 4}.\tag{4.16}$$

For the ρ and a_1 meson we use the spherically symmetric *ansätze*

$$V_\mu = -i\rho_\mu^a \tau^a, \quad \rho_0^a = 0, \quad \rho_i^a = -\epsilon^{aik}\hat{r}^k G(r),$$
$$A_\mu = -ia_\mu^a \tau^a, \quad a_0^a = 0, \quad a_i^a = \hat{r}^i \hat{r}^a F(r) + \delta^{ia} H(r)\tag{4.17}$$

where the index a runs from 1 to 3.

Thus the Euclidian Dirac Hamiltonian reads

$$h = \alpha \cdot p + i\omega(r) + \phi(r)\beta(\cos\Theta(r) + i\gamma_5\tau \cdot \hat{r}\sin\Theta)$$
$$+ \frac{1}{2}(\alpha \times \hat{r})\tau G(r) + \frac{1}{2}(\sigma\cdot\hat{r})(\tau\cdot\hat{r})F(r) + \frac{1}{2}(\sigma \cdot \tau)H(r).\tag{4.18}$$

which obviously is not Hermitean since $\omega(r)$ is real giving rise to complex eigenvalues of h. For static configurations the eigenvalues of ∂_τ, $i\Omega_n = i(2n + 1)\pi/T$, $(n = 0, \pm 1, \pm 2, ..)$ may be separated. Note that the eigenfunctions of ∂_τ assume anti–periodic boundary conditions in the Euclidian time interval T. (The Ω_n are analogous to the Matsubara frequencies with T figuring as inverse temperature.) Thus the eigenvalues $\lambda_{n,\nu}$ of the operator $\partial_\tau + h$ are decomposed according to

$$\lambda_{n,\nu} = -i\Omega_n + \epsilon_\nu = -i\Omega_n + \epsilon_\nu^R + i\epsilon_\nu^I.\tag{4.19}$$

The fermion determinant is expressed in terms of the eigenvalues $\lambda_{n,\nu}$,

$$A_R = \frac{1}{2}\sum_{\nu,n}\log(\lambda_{n,\nu}\lambda_{n,\nu}^*) \quad \text{and} \quad A_I = \frac{1}{2}\sum_{\nu,n}\log(\frac{\lambda_{n,\nu}}{\lambda_{n,\nu}^*}).\tag{4.20}$$

Using (4.19) the real part is given by

$$A_R = \frac{1}{2}\sum_{\nu,n}\log((\Omega_n - \epsilon_\nu^I)^2 + (\epsilon_\nu^R)^2)$$
$$\rightarrow -\frac{1}{2}\sum_{\nu,n}\int_1^\infty \frac{d\tau}{\tau}\exp\{-\frac{\tau}{\Lambda^2}((\Omega_n - \epsilon_\nu^I)^2 + (\epsilon_\nu^R)^2)\}\tag{4.21}$$

where the last line follows in the proper–time regularization scheme. For large Euclidean time intervals ($T \rightarrow \infty$) the temporal part of the trace may be performed

$$A_R = -\frac{T}{2}\sum_\nu \int_{-\infty}^\infty \frac{dz}{2\pi}\int_1^\infty \frac{d\tau}{\tau}\exp\{-\frac{\tau}{\Lambda^2}(z^2 + (\epsilon_\nu^R)^2)\}\tag{4.22}$$

where we have shifted the integration variable $z - \epsilon_\nu^I \to z$. For $T \to \infty$ we may read off the Dirac sea contribution to the real part of the energy functional from $\mathcal{A}_R \to -TE_{\text{vac}}^R$ [Re89]

$$
E_{\text{vac}}^R = \frac{N_c}{4\sqrt{\pi}} \sum_\nu |\epsilon_\nu^R| \Gamma\left(-\frac{1}{2}, (\epsilon_\nu^R/\Lambda)^2\right)
$$

$$
= N_c \sum_\nu \left(\frac{\Lambda}{2\sqrt{\pi}} \exp\left(-(\epsilon_\nu^R/\Lambda)^2\right) + |\epsilon_\nu^R| \mathcal{N}_\nu\right) \tag{4.23}
$$

where

$$
\mathcal{N}_\nu = \frac{1}{\sqrt{\pi}} \Gamma\left(\frac{1}{2}, (\epsilon_\nu^R/\Lambda)^2\right) = -\frac{1}{2}\text{erfc}\left(\left|\frac{\epsilon_\mu}{\Lambda}\right|\right) \tag{4.24}
$$

are the vacuum "occupation numbers" in the proper time regularization scheme. For soliton configurations with vanishing ω (i.e. $\epsilon_\nu^R = \epsilon_\nu$) there is no contribution from the imaginary part and eq. (4.23) is the expression for the energy of the Dirac sea.

For the imaginary part we obtain

$$
\mathcal{A}_I = \frac{1}{2}\left(\sum_\nu \sum_{n=-\infty}^{\infty} \log(\lambda_{\nu,n}) - \sum_\nu \sum_{n=-\infty}^{\infty} \log(\lambda_{\nu,n}^*)\right) = \frac{1}{2}\sum_\nu \sum_{n=-\infty}^{\infty} \log\frac{i\Omega_n - \epsilon_\nu}{i\Omega_n - \epsilon_\nu^*} \tag{4.25}
$$

where we have reversed the sign in the first sum over the integer variable n. Next we express $\mathcal{A}_I$ in terms of a parameter integral,

$$
\mathcal{A}_I = \frac{1}{2}\sum_\nu \sum_{n=-\infty}^{\infty} \int_{-1}^{1} d\lambda \frac{-i\epsilon_\nu^I}{i\Omega_n - \epsilon_\nu^R - i\lambda\epsilon_\nu^I}. \tag{4.26}
$$

In analogy to (4.22) we may carry out the temporal trace in the limit $T \to \infty$:

$$
\mathcal{A}_I = \frac{-i}{2}\sum_\nu \int_{-1}^{1} d\lambda\, T \int_{-\infty}^{\infty} \frac{dz}{2\pi} \epsilon_\nu^I \left[i(z - \lambda\epsilon_\nu^I) - \epsilon_\nu^R\right]^{-1}. \tag{4.27}
$$

Shifting the integration variable $z - \lambda\epsilon_\nu^I \to z$ the integral over λ may be done

$$
\mathcal{A}_I = \frac{-i}{2}\sum_\nu \epsilon_\nu^I \int_{-\infty}^{\infty} \frac{dz}{2\pi} \frac{-2\epsilon_\nu^R}{z^2 + (\epsilon_\nu^R)^2}. \tag{4.28}
$$

$\mathcal{A}_I$ can be regularized in proper time by expressing the integrand as [ZARW94]

$$
\frac{-1}{z^2 + (\epsilon_\nu^R)^2} \to \int_{1/\Lambda^2}^{\infty} d\tau \exp\left\{-\tau(z^2 + (\epsilon_\nu^R)^2)\right\} \tag{4.29}
$$

which obviously is finite for $\Lambda \to \infty$. For simplicity, however, we will treat only the case that $\mathcal{A}_I$ is not regularized. After the z–integration we find for the contribution of the Dirac sea to the imaginary part of the Euclidian energy E_{vac}^I

$$E_{\text{vac}}^I = \frac{-N_c}{2} \sum_\nu \epsilon_\nu^I \text{sign}(\epsilon_\nu^R). \tag{4.30}$$

The total energy functional contains besides E_{vac}^R and E_{vac}^I also the valence quark energy

$$E_{\text{val}}^R = N_c \sum_\nu \eta_\nu |\epsilon_\nu^R| \qquad E_{\text{val}}^I = N_c \sum_\nu \eta_\nu \text{sign}(\epsilon_\nu^R)\epsilon_\nu^I \tag{4.31}$$

with $\eta_\mu = 0, 1$ being the occupation numbers of the valence quark and anti–quark states. Furthermore the meson energy is obtained by substituting the *ansätze* (4.14 – 4.17) into the expression for $\mathcal{A}_m$

$$E_m = 4\pi \int dr r^2 \Big(\frac{\langle \bar{u}u \rangle}{m - m_0} (\phi^2(r) - m^2) + m_\pi^2 f_\pi^2 (1 - \frac{\phi(r)}{m} \cos \Theta(r))$$

$$+ a f_\pi^2 \big(G^2(r) + \frac{1}{2} F^2(r) + F(r)H(r) + \frac{3}{2} H^2(r) - 2\omega^2(r) \big) \Big) \tag{4.32}$$

where $a = 1 + m_\rho^2/6m^2$ if $\pi - a_1$–mixing is included and $a = m_\rho^2/6m^2$ if it is not. Note that we are working in the isospin limit which implies $m_\omega = m_\rho$. Continuing back to Minkowski space we find for the total energy functional[***]

$$E[\phi, \Theta, \omega, G, F, H] = E_{\text{val}}^R + E_{\text{val}}^I + E_{\text{vac}}^R + E_{\text{vac}}^I + E_m. \tag{4.33}$$

The equations of motion for the meson profiles are obtained by extremizing the static Minkowski energy(4.33). In a generic way we may write

$$0 = \frac{\delta E}{\delta \eta} = \frac{\delta E_m}{\delta \eta} + \sum_{\kappa = R, I} \sum_\mu \frac{\partial (E_{\text{val}}^R + E_{\text{val}}^I + E_{\text{vac}}^R + E_{\text{vac}}^I)}{\partial \epsilon_\mu^\kappa} \frac{\delta \epsilon_\mu^\kappa}{\delta \eta} \tag{4.34}$$

wherein η denotes any of the meson profiles $\phi, \Theta, G, \omega, F$ or H. Since h is not Hermitean (in Euclidean space) we have to distinguish between left and right eigenvectors of h. The corresponding eigenvalue equations read

$$h|\Psi_\nu\rangle = \epsilon_\nu |\Psi_\nu\rangle \quad \langle \tilde{\Psi}_\nu | h = \epsilon_\nu \langle \tilde{\Psi}_\nu | \qquad i.e. \ h^\dagger |\tilde{\Psi}_\nu\rangle = \epsilon_\nu^* |\tilde{\Psi}_\nu\rangle. \tag{4.35}$$

The normalization condition is $\langle \tilde{\Psi}_\mu | \Psi_\nu \rangle = \delta_{\mu\nu}$. In order to evaluate the derivatives $\delta \epsilon_\mu^\kappa / \delta \phi$ it is helpful to decompose the Hamiltonian operator (4.18) into Hermitean and anti-Hermitean parts

$$h = h_\Theta + i\omega \tag{4.36}$$

where h_Θ includes all Hermitean terms of the Euclidean Dirac Hamiltonian (4.18). Obviously both, h_Θ and ω, are Hermitean implying $|\tilde{\Psi}_\nu\rangle = |\Psi_\nu^*\rangle$. We

---

[***] Due to some subtilities this expression cannot be considered to be derived rigorously, see ref. [WZAR95].

may therefore extract the real and imaginary parts of the one particle energy eigenvalue

$$\epsilon_\nu^R = \frac{1}{2}\left(\langle\Psi_\nu^*|h|\Psi_\nu\rangle + \langle\Psi_\nu|h^\dagger|\Psi_\nu^*\rangle\right)$$
$$= \langle\Psi_\nu^R|h_\Theta|\Psi_\nu^R\rangle - \langle\Psi_\nu^I|h_\Theta|\Psi_\nu^I\rangle - \langle\Psi_\nu^I|\omega|\Psi_\nu^R\rangle - \langle\Psi_\nu^R|\omega|\Psi_\nu^I\rangle,$$
$$\epsilon_\nu^I = \frac{1}{2}\left(\langle\Psi_\nu^*|h|\Psi_\nu\rangle - \langle\Psi_\nu|h^\dagger|\Psi_\nu^*\rangle\right)$$
$$= \langle\Psi_\nu^R|\omega|\Psi_\nu^R\rangle - \langle\Psi_\nu^I|\omega|\Psi_\nu^I\rangle + \langle\Psi_\nu^I|h_\Theta|\Psi_\nu^R\rangle + \langle\Psi_\nu^R|h_\Theta|\Psi_\nu^I\rangle \qquad (4.37)$$

where we employed the decomposition $|\Psi_\nu\rangle = |\Psi_\nu^R\rangle + i|\Psi_\nu^I\rangle$. Note also that $\langle\Psi_\nu^*| = \langle\Psi_\nu^R| + i\langle\Psi_\nu^I|$. We are now equiped with expressions for the real and imaginary parts of the energy eigenvalues ϵ_μ which are suitable to evaluate the derivatives with respect to the meson fields. For example we have

$$\frac{\delta\epsilon_\nu^I}{\delta\omega(r)} = r^2 \int \frac{d\Omega}{4\pi}\left(\langle\mathbf{r}|\Psi_\nu^R\rangle\langle\Psi_\nu^R|\mathbf{r}\rangle - \langle\mathbf{r}|\Psi_\nu^I\rangle\langle\Psi_\nu^I|\mathbf{r}\rangle\right). \qquad (4.38)$$

The expressions for the functional dependence of the energy eigenvalues may now be substituted into the equations of motion (4.34).

The functional derivative of the one particle energies with respect to the fields yields expressions involving the eigenfunctions (4.35). On the other hand, the derivative of the total energy with respect to the one particle energies results in regularization functions for the vacuum part and the occupation number for the valence part. As the eigenfunctions (4.35) occur only in certain combinations in the equations of motion it is convenient to define quark density matrices. These as well as the equations of motion are given in appendix C.

4.4 The Static NJL Soliton

In this section we present the numerical results characterizing the soliton solution containing all possible grand spin zero meson field profiles, *i.e.* ϕ, π, ω, ρ and a_1. These results will then be compared to cases where some of the vector meson fields are switched off in order to examine the effects of various vector meson fields on the soliton. The meson profiles of the soliton solution are determined by iteration. The first self–consistent soliton solutions were obtained in [RW88] where only the chiral field was considered. For numerical details of the full calculation see ref. [AWZ94].

In Tab. 4.1 we display the energy E of the self-consistent soliton solution for constituent quark masses in the range 350MeV$\leq m \leq$500MeV. This table also contains the various contributions to E as they appear in eq. (4.33) for several values of the constituent mass m. The most striking result observed from Tab. (4.1) is the fact that the real part of the energy eigenvalue associated with the valence quark state is negative! *I.e.* the valence quark has joined the Dirac sea and thus the baryon number is completely carried by the polarized Dirac sea. Thus the soliton of the NJL model supports the picture of baryons as topological

Table 4.1. The soliton energy E as well as its Dirac sea and mesonic contributions E_{vac} and E_{m} for different values of the constituent quark mass m. Also shown is the energy of the 'dived' level (ϵ_{val}). (Taken from ref. [ZARW95].)

m (MeV)	350	400	500
E (MeV)	1125	1091	1022
E_{vac}^R (MeV)	2271	1337	883
E_{vac}^I (MeV)	177	206	223
E_{m} (MeV)	-1323	-454	-83
$\epsilon_{\text{val}}^R/m$	-0.28	-0.51	-0.72
$\epsilon_{\text{val}}^I/m$	0.15	0.15	0.12

solitons of meson fields. As a reminder we would like to mention that for $\epsilon_{\text{val}}^R < 0$ the valence quarks' contribution to the energy is already contained in E_{vac}^R and E_{vac}^I. Thus it must not explicitly be added in (4.31).

Table 4.2. The soliton energy for various treatments of the NJL soliton. The meson fields in the first line denote the allowed meson profiles. All numbers are evaluated for a constituent quark mass $m = 350\text{MeV}$.

	π	π, ρ	π, ω	π, ρ, a_1	π, ω, ρ, a_1	$\phi, \pi, \omega, \rho, a_1$
E (MeV)	1214	957	1343	1010	1139	1125
E_{vac}^R (MeV)	561	655	629	615	562	2271
E_{vac}^I (MeV)	0	0	-22	0	189	177
E_{m} (MeV)	0	155	-42	395	389	-1323
$\epsilon_{\text{val}}^R/m$	0.62	0.14	0.50	-0.13	-0.27	-0.28
$\epsilon_{\text{val}}^I/m$	0	0	0.24	0	0.16	0.15

Next we wish to investigate the role of the different meson fields for the NJL soliton. First, one realizes that the influence of the scalar field ϕ is only minor (except a redistribution from vacuum to mesonic energy). Also the scalar field itself does almost not deviate from its VEV,[†] see Fig. 4.3. Therefore we will neglect the dynamics of the scalar field in the following discussion, *i.e.* we constrain the solitonic fields to be on the chiral circle.

Furthermore, we will compare (see Tab. 4.2) the results obtained for the soliton energy in cases with different (axial-)vector mesons incorporated. Obviously, the inclusion of the ω field always increases the soliton energy while the ρ and a_1 lower the energy. Though the latter result is anticipated the former is somewhat surprising since the meson profiles are obtained by extremizing the total

---

[†] This is only true if the ω is also included. Keeping the scalar field and neglecting the ω leads to an unphysical instability, see ref. [WT92]. This instability can be also avoided if higher order terms for the scalar self–interaction are included [WAW93, MRWSGG92] or if the baryon number is regularized [SARW93].

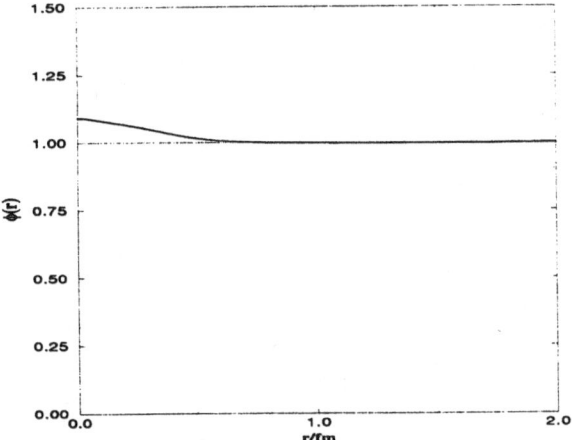

Fig. 4.3. The scalar (chiral radius) field $\Phi(r)$ as a function of the radial distance r for a constituent quark mass $m = 400\mathrm{MeV}$. (Taken from ref. [ZARW95].)

energy and therefore additional degrees of freedom should lower this energy. However, this increase is understood by taking into account that the ω field is proportional to the baryon number density which in turn is constrained by unit baryon number. We also observe from Tab. 4.2 that the inclusion of any of the (axial-) vector mesons lowers the energy eigenvalue of the valence quark. The a_1 obviously effects the valence quark to join the Dirac sea. The ω meson gives a stronger binding of the valence quark. This may be understood by noting that the ω is repulsive yielding a large spatial extension of the soliton. This, in due, causes the valence quark energy to drop.

The repulsive character of the ω field may also be observed directly from the radial behavior of the chiral angle, $\Theta(r)$. In Fig. 4.4 we display $\Theta(r)$ for various treatments of the NJL soliton. $\Theta(r)$ develops the largest tail in the case when the ω meson field is the only one added to the chiral field. The axialvector field provides a significant attraction resulting in a slope for the chiral field which is larger than in the case when $\Theta(r)$ is the only field being present. The inclusion of the ω meson on top of the isovector mesons ρ and a_1 alters the chiral angle only slightly. A similar behavior can be observed for the ρ meson profile $G(r)$ (*cf.* Fig. 4.5) as well as the axialvector meson profiles $F(r)$ and $H(r)$ (*cf.* Fig. 4.6). On the other hand, the inclusion of the isovector mesons on top of the $\pi - \omega$ system significantly reduces the strength of the ω meson profile as may be seen in Fig. 4.7.

In the NJL model the current field identities hold [ER86]. Therefore the axial current $J^i_{5\mu}$ is directly proportional to the axialvector meson field,

$$J^i_{5\mu} = -\frac{1}{4g_2} A^i_\mu, \qquad (4.39)$$

wherein the superscript denotes the isospin component. Noting that g_A is obtained as the matrix element of $J^i_{5\mu}$ at zero momentum transfer we immediately

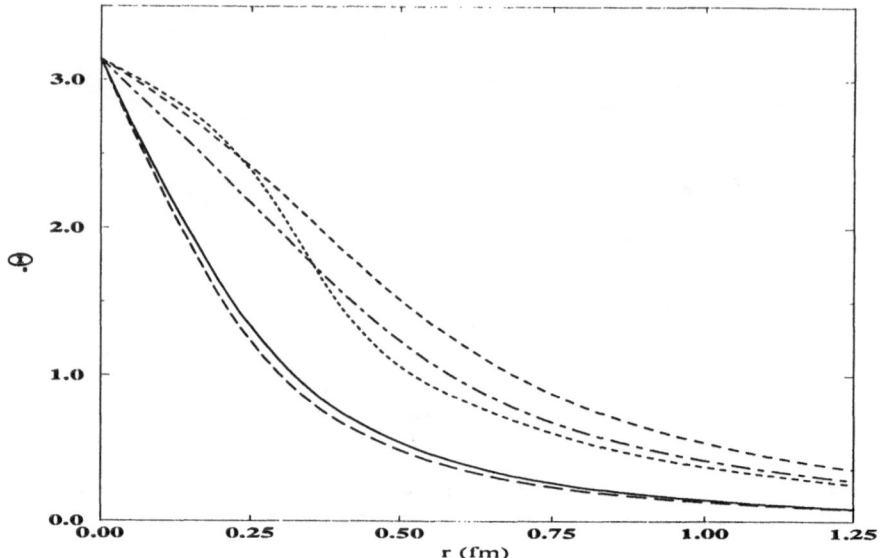

Fig. 4.4. The chiral angle $\Theta(r)$ as a function of the radial distance r. The solid line corresponds to the case when all vector mesons are included; the dashed line to the $\pi - \omega$ system; the short dashed to the $\pi - \rho$ system; the long dashed to the $\pi - \rho - a_1$ system and the dashed dotted to the case when the chiral angle is the only field present. (Taken from ref. [ZARW94].)

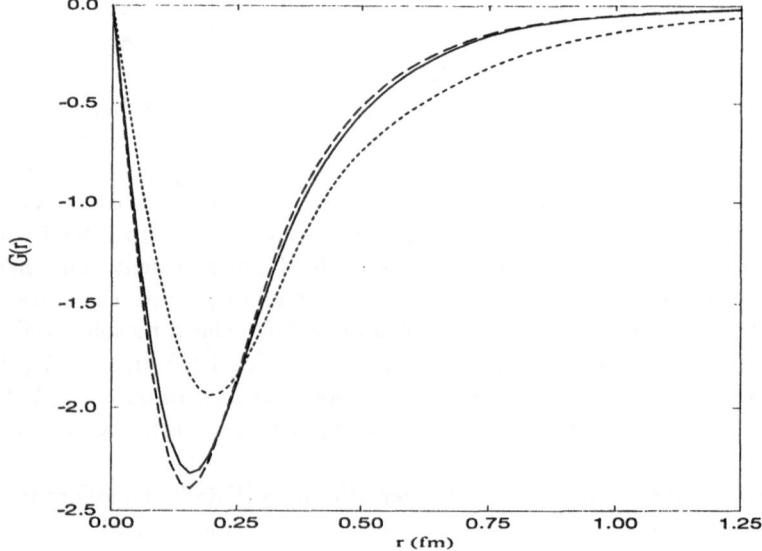

Fig. 4.5. The vector meson profile $G(r)$ as a function of the radial distance r. Solid line: all vector meson fields are present; short dashed denotes the $\pi - \rho$ system and the long dashed to the $\pi - \rho - a_1$ system. (Taken from ref. [ZARW94].)

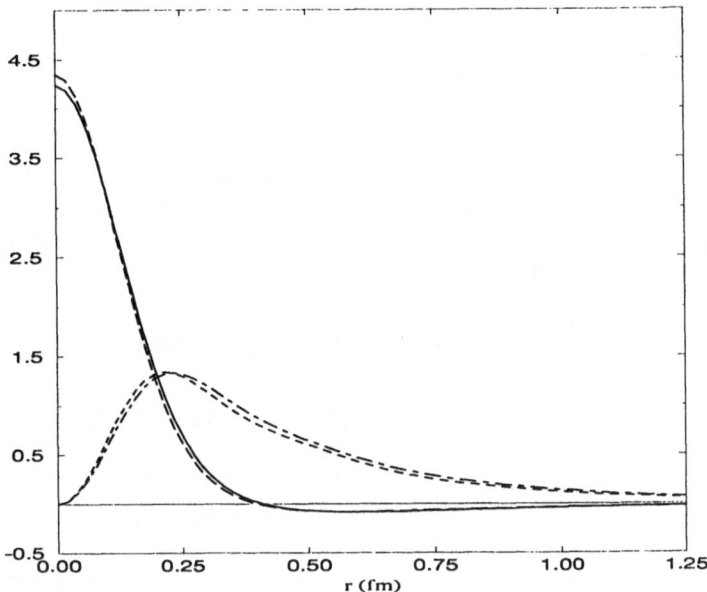

Fig. 4.6. The axialvector meson profile $F(r)$ and $H(r)$ as functions of the radial distance r. $H(r)$ is non-vanishing at the origin. The case when all vector mesons are present is denote by the solid and dashed lines. The $\pi - \rho - a_1$ system is represented by the long dashed and dashed dotted lines. (Taken from ref. [ZARW94].)

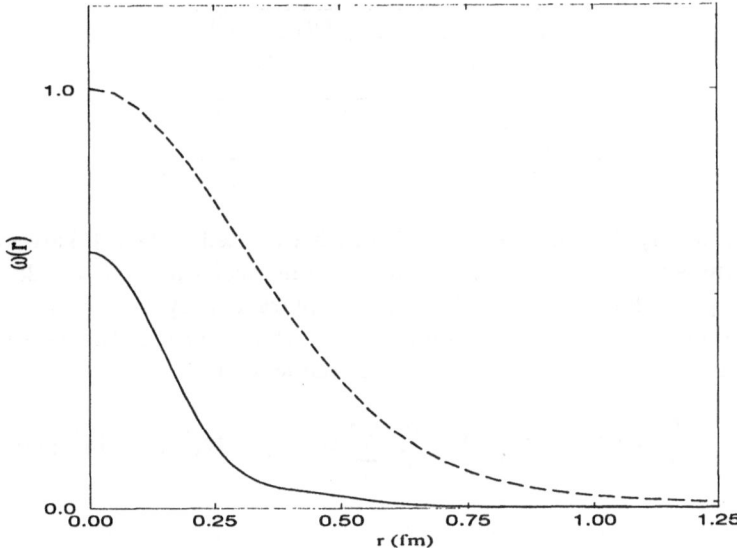

Fig. 4.7. The vector meson profile $\omega(r)$ as a function of the radial distance r. The solid line denotes the case when all vector mesons are present while the dashed line represents the $\pi - \omega$ system. (Taken from ref. [ZARW94].)

obtain [ZARW94]

$$g_A = -\frac{2\pi}{3g_2} \int drr^2 \left[H(r) + \frac{1}{3} F(r) \right] \qquad (4.40)$$

It is important to mention that eq. (4.40) represents an exact result which is not subject to renormalization due to $\pi - a_1$ mixing. Making use of the equation of motion for the axialvector profiles $H(r)$ and $F(r)$ we may reexpress g_A as a mode sum over quark spinors:

$$\begin{aligned}
g_A = -\frac{N_c}{3} \sum_{\mu} \Big\{ &[\ \langle \psi_\mu^R | \sigma_3\tau_3 | \psi_\mu^R \rangle - \langle \psi_\mu^I | \sigma_3\tau_3 | \psi_\mu^I \rangle \\
&+ \langle \psi_\mu^R | \sigma_3\tau_3 | \psi_\mu^I \rangle + \langle \psi_\mu^I | \sigma_3\tau_3 | \psi_\mu^R \rangle \ ] \eta_\mu \\
&+ [\langle \psi_\mu^R | \sigma_3\tau_3 | \psi_\mu^R \rangle - \langle \psi_\mu^I | \sigma_3\tau_3 | \psi_\mu^I \rangle \ ] f_R(\epsilon_\mu/\Lambda) \\
&+ [\langle \psi_\mu^R | \sigma_3\tau_3 | \psi_\mu^I \rangle + \langle \psi_\mu^I | \sigma_3\tau_3 | \psi_\mu^R \rangle \ ] f_I(\epsilon_\mu/\Lambda) \Big\} \qquad (4.41)
\end{aligned}$$

The regulator functions $f_{R,I}$ are given in eqs. (C.3,C.4)). The mode sum (4.41) has the advantage that it may be employed in models without axialvector mesons.

Table 4.3. The axial charge g_A of the nucleon in the various treatments of the soliton in the NJL model.

m (MeV)	300	350	400
π	—	0.78	0.73
π, ω	—	0.98	1.03
π, ρ, a_1	0.31	0.27	0.13
π, ω, ρ, a_1	0.54	0.39	0.28

Unfortunately the numerical results which are listed in Tab. 4.3 are somewhat discouraging since they are well below the numerical value $g_A = 1.25$.[‡] This is especially pronounced in case the valence quark energy is negative. In order to understand the origin of this shortcoming let us consider the case when the isoscalar field ω is absent. Then eq. (4.41) reduces to

$$g_A = -\frac{N_c}{3} \eta_{\text{val}} \langle \text{val} | \sigma_3\tau_3 | \text{val} \rangle + \frac{N_c}{6} \sum_\mu \text{sign}(\epsilon_\mu) \text{erfc}(|\epsilon_\mu|/\Lambda) \langle \mu | \sigma_3\tau_3 | \mu \rangle. \ (4.42)$$

Eq. (4.42) reveals that once the valence quark energy has become negative its contribution to g_A is strongly suppressed by the proper time regularization. This suggests that a regularization prescription which does not affect low-lying states as strongly as the proper time scheme would be highly desirable in order to describe g_A correctly. This consideration is supported by a simple modification of

---

[‡] Also in the Skyrme model g_A comes out too low.

eq. (4.42). In the sum over all eigenstates we replace the complementary error function by a sharp cut-off function. Then the contributions from the low-lying states are not affected by the regularization procedure. Choosing, *e.g.* a constituent mass of 300MeV the prediction for g_A increases drastically to 1.04 in the $\pi - \rho - a_1$ system. Of course, this exploration does not represent a consistent calculation but merely demonstrates that g_A strongly depends on the regularization description. If we consider the $\pi - \omega$ system the strong repulsion of the ω field transfers to an increased prediction for g_A. This is also obvious from Fig. (4.4) since in the absense of axialvector fields g_A may also be obtained from the size of the "pion tail"[ANW83].

We may also investigate the isoscalar radius of the nucleon without explicitly performing the collective quantization (see the subsequent section). Again due to the current field identity the isoscalar charge density is proportional to the ω profile yielding the isoscalar radius

$$\langle r^2_{I=0} \rangle = \frac{4m^2_\rho}{N_c g^2_V} \int d^3r \; r^2 \; \omega(r). \tag{4.43}$$

The resulting data are displayed in Tab. 4.4. The repulsive effect of the isoscalar vector field ω is obvious and is not completely compensated by the attraction provided by the isovector fields ρ and a_1. *I.e.* the prediction still overestimates the experimental value $\langle r^2_{I=0} \rangle^{1/2} \approx 0.8$fm.

Table 4.4. The isoscalar radius $\langle r^2_{I=0} \rangle^{1/2}$ of the nucleon in various treatments of the NJL model. The radii are given in fm.

m (MeV)	300	350	400
π	—	0.89	0.76
π, ρ, a_1	0.55	0.55	0.56
π, ω	—	2.06	1.77
π, ω, ρ, a_1	1.39	1.40	1.29

4.5 Semiclassical Quantization

The chiral hedgehog soliton has neither the good spin nor flavor quantum numbers of the physical baryons. In order to project on baryon quantum numbers for baryons made of u and d quark one employs the cranking procedure [RS80] known from nuclear physics. The main purpose of this section is to explain how semiclassical quantization can be applied to the NJL soliton and furthermore provides a reasonable result for the nucleon-Δ mass difference. Therefore we shall restrict ourselves to the approach which contains pseudoscalar fields only. The semiclassical quantization of the NJL soliton was performed in [Re89]. Below we follow this reference. We start by imposing the ansatz

$$\xi(\boldsymbol{x}, t) = R(t)\xi(\boldsymbol{x})R^{\dagger}(t) \tag{4.44}$$

on the chiral field $U = \xi^2$. $R(t)$ describes hereby the (adiabatic) isorotations.

It is convenient to transform the fermion determinant to the flavor rotating frame $q(\boldsymbol{x}, t) = R(t)q'(\boldsymbol{x}, t)$. Obviously this transformation cancels the rotation matrices in our *ansatz* (4.44) at the expense of induced terms in the quark Hamiltonian due to the time dependence of the rotations

$$h = h_0 + h_{rot} \tag{4.45}$$

wherein h_0 is the static quark Hamiltonian. The induced Hamiltonian h_{rot} originates from the time dependence of the collective rotation

$$h_{rot} = -iR^{\dagger}(t)\dot{R}(t) = \frac{1}{2}\tau\Omega. \tag{4.46}$$

Our goal is to expand the energy in the angular velocities Ω_a. These actually corresponds to an expansion in $1/N_c$. In this expansion we keep terms quadratic in the perturbation h_{rot}. Then, due to isospin invariance, the collective Lagrangian is given by

$$L = -E + \frac{1}{2}\sum_{a,b=1}^{3}\Omega_a\Theta_{ab}\Omega_b = -E + \frac{1}{2}\alpha^2\Omega^2 \tag{4.47}$$

where E is the classical soliton energy. Isospin invariance demands that the matrix for the moment of inertia, Θ_{ab}, is diagonal and only one independent component, Θ_{33}, corresponding to α^2 exist. The valence quark contribution to the expressions quadratic in angular velocities, $\frac{1}{2}\Theta_{ab}\Omega_a\Omega_b$, is the well known cranking result

$$\Theta_{ab}^{val} = \frac{N_c}{2}\eta^{val}\sum_{\mu\neq val}\frac{\langle val|\tau_a|\mu\rangle\langle\mu|\tau_b|val\rangle}{\epsilon_\mu - \epsilon_{val}}. \tag{4.48}$$

Next we turn to the evaluation of the vacuum contribution to the fermion determinant in presence of the perturbation h_{rot}. As for the calculation of the static soliton solution we transform to Euclidean space and take the vacuum ($\xi = 1$) as reference

$$\mathcal{A}_F = \mathrm{Tr}\log\left(i\rlap{/}{D}(\xi(\boldsymbol{x}, t))\right) - \mathrm{Tr}\log\left(i\rlap{/}{D}(\xi = 1)\right). \tag{4.49}$$

In the flavor rotating frame we have

$$\bar{q}\rlap{/}{D}q = \bar{q}'\rlap{/}{D}'q', \qquad \text{with} \qquad i\rlap{/}{D}' = \beta\left(\partial_\tau - (h_0 + h_{rot})\right). \tag{4.50}$$

Attention has to be paid to the fact that in Euclidean space the angular velocities Ω_a are to be considered anti-Hermitean quantities.

It is sufficient to consider the real part of the action because for isorotations the imaginary part vanishes. The real part is in the flavor rotating frame given by

$$A_R = -\frac{1}{2} \int_{1/\Lambda^2}^{\infty} \frac{ds}{s} \operatorname{Tr} \exp(-s\not{D}'^{\dagger}\not{D}'). \tag{4.51}$$

This quantity is expanded up to second order in Ω. (The underlying assumption of all these calculations is that of a strictly adiabatic rotation.) Then we find for the moments of inertia

$$\Theta_{ab}^{\text{vac}} = \frac{N_c}{4} \sum_{\mu\nu} f_\Theta(\epsilon_\mu, \epsilon_\nu; \Lambda) \langle \mu | \tau_a | \nu \rangle \langle \nu | \tau_b | \mu \rangle \tag{4.52}$$

where the cut-off function $f_\Theta(\epsilon_\mu, \epsilon_\nu; \Lambda)$ is given by

$$f_\Theta(\epsilon_\mu, \epsilon_\nu; \Lambda) = \frac{\Lambda}{\sqrt{\pi}} \frac{e^{-\epsilon_\mu^2/\Lambda^2} - e^{-\epsilon_\nu^2/\Lambda^2}}{\epsilon_\nu^2 - \epsilon_\mu^2} - \frac{2}{\epsilon_\mu - \epsilon_\nu} \left[\operatorname{sign}(\epsilon_\nu)\mathcal{N}_\mu - \operatorname{sign}(\epsilon_\mu)\mathcal{N}_\nu \right]. \tag{4.53}$$

$\mathcal{N}_\mu$ is the vacuum occupation number defined in eq. (4.24).

The Hamiltonian operator

$$H = -\sum_{a=1}^{3} J_a \Omega_a - L \tag{4.54}$$

may be diagonalized yielding the baryon energy (mass)

$$E_B = E + \frac{1}{2\alpha^2} J(J+1). \tag{4.55}$$

Substituting now the eigenvalues of J for the nucleon ($J = 1/2$) and the Δ ($J = 3/2$) the nucleon–Δ mass difference is calculated to be

$$M_\Delta - M_N = \frac{3}{2\alpha^2}. \tag{4.56}$$

For a constituent mass $m = 400\text{MeV}$ the numerical calculation provides a value $M_\Delta - M_N = 296\text{MeV}$ which has to be compared with the experimental value 293MeV. Constituent masses of 350 or 450 MeV provide a mass splitting of 209 or 350 MeV, respectively. Thus, one obtains reasonable values for the nucleon–Δ mass difference if reasonable values for the constituent mass are used.

4.6 Meson Fluctuations off the Soliton

In this section we will consider mesonic fluctuations in the soliton background [WRA93]. For pedagogical reasons we will restrict ourselves to the discussion of pion modes in the isospin limit. Especially, this will be used to discuss the zero modes of the soliton in the next section.

We parametrize the matrix M in a way which will prove appropriate for discussing fluctuations off the soliton

$$M = \xi_0 \xi_f \Sigma \xi_f \xi_0. \tag{4.57}$$

The matrix Σ is hermitian while ξ_0 and ξ_f are unitary. Note that this parametrization differs from the usually adopted unitary gauge $M = \xi_L^\dagger \Sigma \xi_R$, $\xi_L^\dagger = \xi_R$. In the baryon number zero sector ξ_0 is replaced by the unit matrix. Space–time dependent fluctuating pseudoscalar meson fields $\pi_i(x)$ are introduced via (see also eq. (3.50))

$$\xi_f(x) = \exp\left(i \sum_{i=1}^{3} \pi_i(x)\tau_i/2 \right). \tag{4.58}$$

For the ongoing calculations involving the static chiral soliton $\xi_0(\boldsymbol{r})$ the meson fields will be constrained to the chiral circle. We thus write for the matrix M

$$M = \xi_0 \xi_f \langle \Sigma \rangle \xi_f \xi_0.$$

$$\xi_0(x) = \exp\left(\frac{i}{2}\boldsymbol{\tau} \cdot \hat{\boldsymbol{r}}\, \Theta(r) \right). \tag{4.59}$$

The main goal of our calculations is to expand the action $\mathcal{A}$ in the presence of the soliton (4.59) up to second order in the space-time dependent meson fluctuations $\pi_i(x)$. First, we express the Euclidean Dirac operator $\not{D}$ as

$$i\beta\not{D} = -\partial_\tau - \boldsymbol{\alpha} \cdot \boldsymbol{p} - T\beta \left(\xi_f \langle \Sigma \rangle \xi_f P_R + \xi_f^\dagger \langle \Sigma \rangle \xi_f^\dagger P_L \right) T^\dagger \tag{4.60}$$

wherein $\tau = ix_0$ is the euclidean time. The unitary matrix

$$T = \xi_0 P_L + \xi_0^\dagger P_R \tag{4.61}$$

contains the information about the chiral soliton. The appearance of the meson fluctuations is most clearly exhibited by introducing a hamiltonian h in (4.60)

$$i\beta\not{D} = -\partial_\tau - h = -\partial_\tau - \left(h_{(0)} + h_{(1)} + h_{(2)} + \cdots \right) \tag{4.62}$$

wherein the subscript labels the power of the meson fluctuations,

$$h_{(0)}(\mathbf{r}) = \boldsymbol{\alpha} \cdot \boldsymbol{p} - T\beta\langle \Sigma \rangle T^\dagger$$

$$= \boldsymbol{\alpha} \cdot \boldsymbol{p} + \beta m(\cos\Theta + i\boldsymbol{\tau} \cdot \hat{\boldsymbol{r}}\gamma_5 \sin\Theta) \tag{4.63}$$

$$h_{(1)}(\mathbf{r}, t) = iT\beta\gamma_5 m\boldsymbol{\pi} \cdot \boldsymbol{\tau} T^\dagger \tag{4.64}$$

$$h_{(2)}(\mathbf{r}, t) = -T\beta\frac{m}{2}\boldsymbol{\pi} \cdot \boldsymbol{\pi} T^\dagger. \tag{4.65}$$

Obviously, the quantities $h_{(i)}$ are hermitian operators. Noting furthermore that $h_{(0)}$ is time independent we have up to second order in the fluctuations

$$\not{D}^{\dagger}\not{D} = -\partial_\tau^2 + h_{(0)}^2 - [\partial_\tau, h_{(1)}] + \{h_{(1)}, h_{(0)}\}$$
$$- [\partial_\tau, h_{(2)}] + \{h_{(2)}, h_{(0)}\} + h_{(1)}^2 + \cdots. \tag{4.66}$$

It is useful to define a zeroth-order heat kernel operator [Re89, WAR92] (see also Appendix A) $\hat{K}_0(s) = \exp\left(s(\partial_\tau^2 - h_{(0)}^2)\right)$ in order to expand the action§

$$\mathcal{A} = \mathcal{A}^{(0)} + \mathcal{A}^{(1)} + \mathcal{A}^{(2)} + \cdots \tag{4.67}$$

$$\mathcal{A}^{(0)} = -\frac{1}{2}\mathrm{Tr}\int_{1/\Lambda^2}^{\infty}\frac{ds}{s}\hat{K}_0(s) \tag{4.68}$$

$$\mathcal{A}^{(1)} = \frac{1}{2}\mathrm{Tr}\int_{1/\Lambda^2}^{\infty}ds\hat{K}_0(s)\{h_{(1)}, h_{(0)}\} \tag{4.69}$$

$$\mathcal{A}^{(2)} = \frac{1}{2}\mathrm{Tr}\int_{1/\Lambda^2}^{\infty}ds\hat{K}_0(s)\left(\{h_{(2)}, h_{(0)}\} + h_{(1)}^2\right) - \frac{1}{4}\mathrm{Tr}\int_{1/\Lambda^2}^{\infty}ds\int_0^s ds'\hat{K}_0(s-s')$$
$$\times \left([\partial_\tau, h_{(1)}]\hat{K}_0(s')[\partial_\tau, h_{(1)}] + \{h_{(1)}, h_{(0)}\}\hat{K}_0(s')\{h_{(1)}, h_{(0)}\}\right). \tag{4.70}$$

Again extensive use has been made of $[\partial_\tau, h_{(0)}] = 0$.

For carrying out the temporal part of the functional trace we perform a Fourier transformation of the meson fluctuations in Euclidean space

$$\pi_i(\mathbf{r}, -i\tau) = \int_{-\infty}^{+\infty}\frac{d\omega}{2\pi}\,\tilde{\pi}_i(\mathbf{r}, i\omega)e^{-i\omega\tau} \tag{4.71}$$

which may be transfered to the Hamiltonians,

$$h_{(1)}(\mathbf{r}, -i\tau) = \int_{-\infty}^{+\infty}\frac{d\omega}{2\pi}\tilde{h}_{(1)}(\mathbf{r}, i\omega)e^{-i\omega\tau} \quad \text{and}$$

$$h_{(2)}(\mathbf{r}, -i\tau) = \int_{-\infty}^{+\infty}\frac{d\omega}{2\pi}\int_{-\infty}^{+\infty}\frac{d\omega'}{2\pi}\tilde{h}_{(2)}(\mathbf{r}, i\omega, i\omega')e^{-i(\omega+\omega')\tau} \tag{4.72}$$

wherein $\tilde{h}_{(i)}$ are obtained from $h_{(i)}$ (4.64,4.65) through substitution of the meson fields by their Fourier transforms (4.71). Note that for actual computations the frequency ω has to be continued back to Minkowski space. The temporal trace essentially boils down to computing gaussian integrals. The spatial part of the trace as well as the traces over Dirac and flavor indices are evaluated using the eigenstates of the static one-particle Hamiltonian $h_{(0)}$

$$h_{(0)}|\mu\rangle = \epsilon_\mu|\mu\rangle. \tag{4.73}$$

§ The imaginary part of the action $\mathcal{A}_I$ does not contribute in the two flavor case. Therefore it is sufficient to consider the real part $\mathcal{A}_R$.

The zeroth order term in this expansion is obviously proportional to the vacuum part of the classical soliton energy

$$\mathcal{A}^{(0)} = -T\frac{N_c}{2} \int_{1/\Lambda^2}^{\infty} \frac{ds}{\sqrt{4\pi s^3}} \sum_{\mu} \exp(-s\epsilon_{\mu}^2) = -TE^{\text{vac}}[\Theta]. \qquad (4.74)$$

We will show below that the linear term

$$\mathcal{A}^{(1)} = N_c \int_{1/\Lambda^2}^{\infty} ds \sum_{\mu} e^{-s\epsilon_{\mu}^2} \epsilon_{\mu} \langle\mu|\tilde{h}_{(1)}(\mathbf{r}, 0)|\mu\rangle \qquad (4.75)$$

is proportional to the vacuum contribution to the equation of motion. This is an expected result because the classical soliton configuration extremizes the action, and therefore the whole linear term has to vanish by virtue of the equation of motion.

The result for the quadratic term is (in Minkowski space)

$$\mathcal{A}^{(2)} = \frac{N_c}{2} \int_{1/\Lambda^2}^{\infty} \frac{ds}{\sqrt{4\pi s}} \sum_{\mu} 2\epsilon_{\mu} e^{-s\epsilon_{\mu}^2} \int_{-\infty}^{+\infty} \frac{d\omega}{2\pi} \langle\mu|\tilde{h}_{(2)}(\mathbf{r}, \omega, -\omega)|\mu\rangle$$

$$+ \frac{N_c}{4} \int_{1/\Lambda^2}^{\infty} ds \sqrt{\frac{s}{4\pi}} \sum_{\mu\nu} \int_{-\infty}^{+\infty} \frac{d\omega}{2\pi} \langle\mu|\tilde{h}_{(1)}(\mathbf{r}, \omega)|\nu\rangle \langle\nu|\tilde{h}_{(1)}(\mathbf{r}, -\omega)|\mu\rangle$$

$$\times \left\{ \frac{e^{-s\epsilon_{\mu}^2} + e^{-s\epsilon_{\nu}^2}}{s} + [\omega^2 - (\epsilon_{\mu} + \epsilon_{\nu})^2]R(s; \omega, \epsilon_{\mu}, \epsilon_{\nu}) \right\}. \qquad (4.76)$$

The regulator function R involves a Feynman parameter integral reflecting the quark loop in the presence of the soliton

$$R(s; \omega, \epsilon_{\mu}, \epsilon_{\nu}) = \int_0^1 dx \, \exp\left(-s[(1-x)\epsilon_{\mu}^2 + x\epsilon_{\nu}^2 - x(1-x)\omega^2]\right). \qquad (4.77)$$

Besides the Dirac sea also the explicit occupation of the valence quark level contributes to the action as long as the corresponding energy eigenvalue ϵ_{val} is positive. Since no regularization is involved the computation can be completely performed in Minkowski space using ordinary perturbation theory. Up to second order in the fluctuations the corresponding contribution to the action reads

$$\mathcal{A}_{\text{val}} = -\eta^{\text{val}} N_c \left\{ T\epsilon_{\text{val}} + \langle\text{val}|\tilde{h}_{(1)}(\mathbf{r}, 0)|\text{val}\rangle + \int_{-\infty}^{+\infty} \frac{d\omega}{2\pi} \left(\langle\text{val}|\tilde{h}_{(2)}(\mathbf{r}, \omega, -\omega)|\text{val}\rangle \right. \right.$$

$$\left. \left. + \sum_{\mu\neq\text{val}} \langle\text{val}|\tilde{h}_{(1)}(\mathbf{r}, \omega)|\mu\rangle \langle\mu|\tilde{h}_{(1)}(\mathbf{r}, -\omega)|\text{val}\rangle \frac{\epsilon_{\text{val}} - \epsilon_{\mu}}{(\epsilon_{\text{val}} - \epsilon_{\mu})^2 - \omega^2} \right) \right\}. $$

$$(4.78)$$

Here $\eta^{\text{val}} = 0, 1$ denotes the occupation number of the valence quark and anti-quark states. To obtain this second order result the expression for the associated first order change $\delta\Psi_{\mathrm{val}}$ of the valence quark wave-function Ψ_{val}

$$\delta\Psi_{\mathrm{val}}(\mathbf{r},t) = \left(i\partial_t - h_{(0)}(\mathbf{r})\right)^{-1} h_{(1)}(\mathbf{r},t)\Psi_{\mathrm{val}}(\mathbf{r},t) \qquad (4.79)$$

has been used.

Finally we need to expand the purely mesonic part of the action $\mathcal{A}_m$ (see eq. (3.48)). This can be done straightforwardly yielding

$$\mathcal{A}_m = -Tm_\pi^2 f_\pi^2 \int d^3r \; (1 - \cos\Theta) - m_\pi^2 f_\pi^2 \int d^3r \; \sin\Theta \; \hat{\mathbf{r}} \cdot \tilde{\boldsymbol{\pi}}(0)$$

$$-\frac{1}{2} m_\pi^2 f_\pi^2 \int d^3r \int_{-\infty}^{+\infty} \frac{d\omega}{2\pi} \cos\Theta \; \tilde{\boldsymbol{\pi}}(\omega) \cdot \tilde{\boldsymbol{\pi}}(-\omega) + \dots \qquad (4.80)$$

wherein we made use of the relation $G_1 = m_0 m / m_\pi^2 f_\pi^2$ [ER86].

As promised above we will prove that the linear term vanishes if the soliton fulfills the equation of motion. The zeroth-order hamiltonian $h_{(0)}$ commutes with the grand spin operator $\mathbf{G} = \mathbf{l} + \boldsymbol{\sigma}/2 + \boldsymbol{\tau}/2$. Thus the eigenstates $|\mu\rangle$ in (4.73) are degenerate with respect to the grand spin projection quantum number M_G. This implies that the sum over M_G in (4.75) and in the linear term of eq. (4.78) projects out the grand spin zero piece of $\tilde{h}_{(1)}$

$$\hat{P}_{G=0}\left(\tilde{h}_{(1)}(\mathbf{r},0)\right) = \hat{m}\beta\left(-\sin\Theta + i\boldsymbol{\tau} \cdot \hat{\mathbf{r}} \; \gamma_5\cos\Theta\right) \hat{P}_{L=0}\left(\hat{\mathbf{r}} \cdot \tilde{\boldsymbol{\pi}}(\mathbf{r},0)\right). \qquad (4.81)$$

$\hat{P}_{G=0}\left(\tilde{h}_{(1)}(\mathbf{r},0)\right)$ is even under parity transformations since $\hat{P}_{L=0}\left(\hat{\mathbf{r}} \cdot \tilde{\boldsymbol{\pi}}(\mathbf{r},0)\right)$ reduces to a radial function. Noting furthermore that $\tilde{\boldsymbol{\pi}}(\mathbf{r},0) = \int dt \, \boldsymbol{\pi}(\mathbf{r},t)$, the linear piece of the action collects up to

$$\mathcal{A}^{(1)} = N_c m \int d^4x \; \hat{P}_{L=0}\left(\hat{\mathbf{r}} \cdot \boldsymbol{\pi}(\mathbf{r},t)\right)$$

$$\times\left\{\sin\Theta\left[-\frac{m_\pi^2 f_\pi^2}{N_c m} + \mathrm{tr}\left(\sum_\mu \mathrm{sign}(\epsilon_\mu)\mathcal{N}_\mu\Psi_\mu(\mathbf{r})\bar{\Psi}_\mu(\mathbf{r}) + \eta^{\mathrm{val}}\Psi_{\mathrm{val}}(\mathbf{r})\bar{\Psi}_{\mathrm{val}}(\mathbf{r})\right)\right]\right.$$

$$\left. -\cos\Theta \; \mathrm{tr}\left[i\boldsymbol{\tau} \cdot \hat{\mathbf{r}} \; \gamma_5\left(\sum_\mu \mathrm{sign}(\epsilon_\mu)\mathcal{N}_\mu\Psi_\mu(\mathbf{r})\bar{\Psi}_\mu(\mathbf{r}) + \eta^{\mathrm{val}}\Psi_{\mathrm{val}}(\mathbf{r})\bar{\Psi}_{\mathrm{val}}(\mathbf{r})\right)\right]\right\}.$$

$$(4.82)$$

Here the trace runs over Dirac and isospin indices only. $\mathcal{N}_\mu$ refers to the vacuum occupation numbers in the proper time regularization scheme, see eq. (4.24). The expression in curly brackets in eq. (4.82) may easily be verified to be the equation of motion for the chiral angle [Re89, WAR92], see Appendix C. Thus the linear term vanishes for the selfconsistent chiral soliton. This proves that the hedgehog represents a local extremum of the NJL action. The main ingredient of this proof is the completeness of the basis (4.73), $i.e.$ the fact that the functional trace may be computed using the eigenstates of $h_{(0)}$.

4.7 Zero Modes

A special type of fluctuation is given by the zero modes, *i.e.* fluctuations off the soliton with zero frequency. One can extract them using symmetry considerations. They arise whenever the background field (the "vacuum" seen by the meson fluctuations) breaks a symmetry of the underlying theory. In the case of the chiral soliton the hedgehog field configuration violates the rotational and translational invariance. Zero modes are therefore associated with infinitesimal spatial rotations and translations of the soliton. Due to the grand spin symmetry the zero mode corresponding to infinitesimal iso–rotations is equivalent to the one of the spatial–rotations. Although the model is invariant under axial rotations (for massless pions) a corresponding zero mode does not exist since the infinitesimal axial rotation does not leave the vacuum configuration $\langle \Sigma \rangle = m$ invariant.

In order to identify the zero modes we expand the pion field, *i.e.* hedgehog and fluctuations, see eq. (4.59), up to linear order in the fluctuation

$$M = m \left\{ \xi_0^2 + i\xi_0 \boldsymbol{\pi} \cdot \boldsymbol{\tau}\xi_0 \right\} + \dots \tag{4.83}$$

For the extraction of the formal structure of the zero modes we have to identify the linear term with $[\mathcal{G}, M_0]$. Here $\mathcal{G}$ refers to the generator of the symmetry transformation. For the spatial rotation this gives

$$\boldsymbol{\pi}_R(\boldsymbol{r}) = \sin\Theta(r)\hat{\boldsymbol{r}} \times \boldsymbol{\delta}_R \tag{4.84}$$

where $\boldsymbol{\delta}_R$ is a measure for the infinitesimal rotation. Similarly the translation defines

$$\boldsymbol{\pi}_T(\boldsymbol{r}) = \Theta'(r)\hat{\boldsymbol{r}}\,\hat{\boldsymbol{r}} \cdot \boldsymbol{\delta}_T + \frac{\sin\Theta(r)}{r}\left(\boldsymbol{\delta}_T - \hat{\boldsymbol{r}}\,\hat{\boldsymbol{r}} \cdot \boldsymbol{\delta}_T\right). \tag{4.85}$$

It is straightforward to verify that both $\boldsymbol{\pi}_R$ and $\boldsymbol{\pi}_T$ carry unit grand spin, *i.e.* they are dipoles in grand spin space. These channels are usually called magnetic dipole ($M1$) for the rotational zero mode and electric dipole ($E1$) for the translational zero mode, respectively.

On the other hand, the zero modes obviously have also to be found by extremizing the second order part of the action $\mathcal{A}^{(2)}$ with respect to the fluctuation π. In a first step we express $\mathcal{A}^{(2)}$ with the help of a local and a bilocal kernel, $\Phi_1^{ab}(\boldsymbol{r})$ and $\Phi_2^{ab}(\boldsymbol{r}, \boldsymbol{r}', \omega)$ respectively,

$$\mathcal{A}_2 = \frac{1}{2} \int_{-\infty}^{+\infty} \frac{d\omega}{2\pi} \left\{ \int d^3r \int d^3r' \Phi_2^{ab}(\boldsymbol{r}, \boldsymbol{r}', \omega)\tilde{\pi}_a(\boldsymbol{r}, \omega)\tilde{\pi}_b(\boldsymbol{r}', -\omega) \right.$$

$$\left. + \int d^3r \Phi_1^{ab}(\boldsymbol{r})\tilde{\pi}_a(\boldsymbol{r}, \omega)\tilde{\pi}_b(\boldsymbol{r}, -\omega) \right\}. \tag{4.86}$$

These kernels can be extracted from the expressions given in the last section. The local kernel turns out to be diagonal in isospace $\Phi_1^{ab}(\boldsymbol{r}) = \Phi_1(r)\delta^{ab}$. Using the above defined unitary matrix T (4.61) the presentation of these kernels can be considerably simplified by defining chirally rotated wave–functions $\tilde{\Psi}_\mu = T^\dagger \Psi_\mu$. The local kernel is found to be independent of the frequency,

$$\Phi_1(\boldsymbol{r}) = -m_\pi^2 f_\pi^2 \cos\Theta(r) + 2\eta_{\text{val}} N_c m \tilde{\Psi}_{\text{val}}^\dagger(\boldsymbol{r}) \beta \tilde{\Psi}_{\text{val}}(\boldsymbol{r})$$

$$-2N_c m \int_{1/\Lambda^2}^\infty \frac{ds}{\sqrt{4\pi s}} \sum_\mu \epsilon_\mu e^{-s\epsilon_\mu^2} \tilde{\Psi}_\mu^\dagger(\boldsymbol{r}) \beta \tilde{\Psi}_\mu(\boldsymbol{r}).$$

$$(4.87)$$

The bilocal kernel is given by a more complicated expression

$$\Phi_2^{ab}(\boldsymbol{r}, \boldsymbol{r}', \omega) =$$

$$2\eta_{\text{val}} N_c m^2 \sum_{\mu \neq \text{val}} \tilde{\Psi}_{\text{val}}^\dagger(\boldsymbol{r}) \beta\gamma_5 \tau^a \tilde{\Psi}_\mu(\boldsymbol{r}) \tilde{\Psi}_\mu^\dagger(\boldsymbol{r}') \beta\gamma_5 \tau^b \tilde{\Psi}_{\text{val}}(\boldsymbol{r}') \frac{\epsilon_{\text{val}} - \epsilon_\mu}{(\epsilon_{\text{val}} - \epsilon_\mu)^2 - \omega^2}$$

$$-\frac{N_c}{2} m^2 \int_{1/\Lambda^2}^\infty ds \sqrt{\frac{s}{4\pi}} \sum_{\mu\nu} \tilde{\Psi}_\nu^\dagger(\boldsymbol{r}) \beta\gamma_5 \tau^a \tilde{\Psi}_\mu(\boldsymbol{r}) \tilde{\Psi}_\mu^\dagger(\boldsymbol{r}') \beta\gamma_5 \tau^b \tilde{\Psi}_\nu(\boldsymbol{r}')$$

$$\times \left\{ \frac{e^{-s\epsilon_\mu^2} + e^{-s\epsilon_\nu^2}}{s} + [\omega^2 - (\epsilon_\mu + \epsilon_\nu)^2] R(s; \omega, \epsilon_\mu, \epsilon_\nu) \right\} \qquad (4.88)$$

where the function R is defined in eq. (4.77). These kernels are diagonal in parity and grand spin, *i.e.* meson fluctuations with different grand spin and/or parity quantum numbers decouple. This is also true for non–zero modes and is a consequence of the fact that the classical soliton carries grand spin zero and has definite parity. In the zero mode channels this decoupling is very helpful because it allows us to consider rotational and translational zero modes separately.

The Bethe–Salpeter equation for the pion fluctuations in the soliton background, *i.e.* the equation of motion for the fluctuations, is finally obtained by varying (4.86) with respect to the fluctuation π_a

$$\int d^3r' \, \Phi_2^{ab}(\boldsymbol{r}, \boldsymbol{r}', \omega) \tilde{\pi}_b(\boldsymbol{r}', \omega) + \Phi_1(\boldsymbol{r}) \tilde{\pi}_a(\boldsymbol{r}, \omega) = 0. \qquad (4.89)$$

This linear integral equation constitutes an eigenvalue problem. The frequency ω has to be adjusted to a certain value ω_n such that (4.89) is satisfied for a non–trivial $\tilde{\pi}_a(\boldsymbol{r}, \omega_n)$. The frequency ω_n is then the eigenfrequency and the corresponding $\tilde{\pi}_a(\boldsymbol{r}, \omega_n)$ is the associated eigenmode or wavefunction.

In order to numerically solve eq. (4.89) one has to discretize the radial coordinate thereby generating a matrix equation. In the general case this matrix equation has to be generalized to an eigenvalue problem by setting the right hand side to $\lambda(\omega)\pi_a(\boldsymbol{r}, \omega)$. The eigenvalue $\lambda(\omega)$ then depends on the frequency ω. Finally, ω is tuned to the eigenfrequency ω_i such that $\lambda(\omega_i) = 0$. The associated eigenvector represents the eigenmode in the descretized form. To obtain sufficient accuracy in such a procedure is far from trivial. However, for the zero modes we already know the eigenfrequency and a much simpler method can

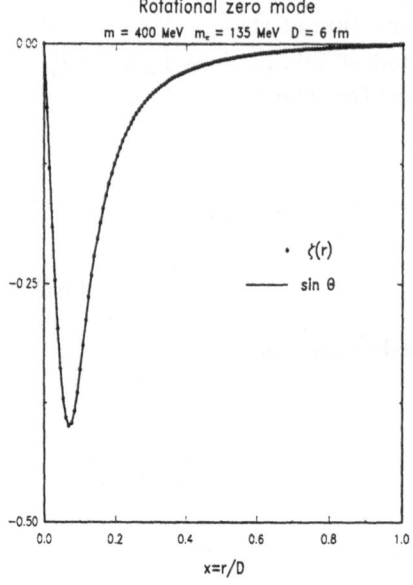
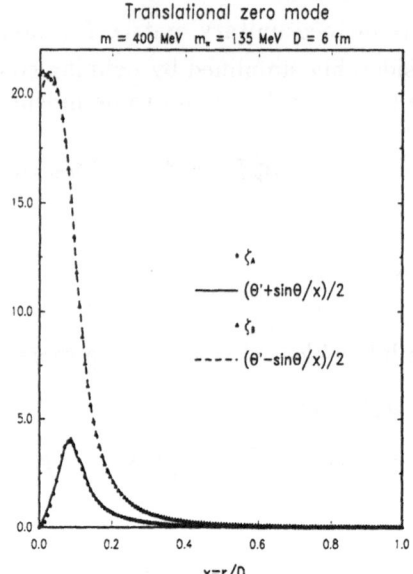

Fig. 4.8. Comparison of the numerical solution of the Bethe–Salpeter equations with the analytic form for the zero modes (taken from ref. [WAR95]).

be used. One diagonalizes the discretized kernels for $\omega_i = 0$ and extracts the wavefunctions which is the one associated with the lowest eigenvalue.

Now let us compare numerical results for the Bethe–Salpeter equation with the analytical expressions from the symmetry considerations. The results are shown in Fig. 4.8 for the constituent quark mass $m = 400\,\text{MeV}$. For the rotational zero mode we find excellent agreement, while for the translational zero mode a small deviation can be observed in the vicinity of the origin due to numerical problems. Thus the existence of zero modes in the background field of the NJL soliton is proven numerically.

4.8 Quantum Corrections to the Soliton Mass

It is quite obvious that the classical energy of the soliton does not yet represent baryon masses. We have already seen in the section on semiclassical quantization that there are quantum contributions. As a matter of fact, these corrections, which are due to mesonic loops, can be classified by the $1/N_c$–expansion. As the absolute mass of the nucleon (938MeV) is in leading order proportional to N_c, the quantum corrections are of one order less, N_c^0. The third order in this expansion (N_c^{-1}) is then the cranking contribution leading to the nucleon–Δ mass splitting. In this section we will be concerned with the order one corrections.

It turns out that these quantum corrections are negative and that the dominant contribution is due to the zero modes. We will not give here the quite complicated derivation of these quantities but refer the reader to the literature [WAR95]. The general idea is that quantum corrections are naturally subject to subtracting the counterpart associated with the trivial vacuum. However, in the trivial vacuum these zero modes are not present, on the contrary there is a mass gap in these channels. From these considerations it becomes obvious that the quantum corrections to the soliton mass are negative. We also understand why the zero modes play such a prominent role: Other modes are much less effected by the presence of the soliton, and therefore the direct contribution and the vacuum subtraction cancel quite accurately.

Table 4.5. The quantum corrections to the soliton mass due to the rotational zero mode (taken from ref. [WAR95]).

	$m_\pi = 0$			$m_\pi = 135\text{MeV}$		
$m(\text{MeV})$	400	500	600	400	500	600
$\Delta E(\text{MeV})$	-201	-274	-290	-244	-297	-323

Table 4.6. The quantum corrections to the soliton mass due to the translational zero mode. The contributions stemming from the $S(l = 0)$- and $D(l = 2)$-waves are disentangled (taken from ref. [WAR95]).

	$m_\pi = 0$			$m_\pi = 135\text{MeV}$		
$m(\text{MeV})$	400	500	600	400	500	600
$\Delta E_{l=0}(\text{MeV})$	-18	-22	-28	-12	-22	-28
$\Delta E_{l=2}(\text{MeV})$	-127	-140	-207	-82	-128	-187
$\Delta E(\text{MeV})$	-145	-162	-235	-94	-150	-215

In Tabs. 4.5 and 4.6 results for the leading quantum corrections to the NJL soliton are displayed. One sees that the corrections are sizeable, -(250 - 300) MeV due to the rotational zero modes and -(100 - 200) MeV due to translational zero modes. Including these corrections, *i.e.* the sum of the zero mode contributions listed in Tabs. 4.5 and 4.6, as well as the cranking contribution to the baryon masses,

$$M_{\text{baryon}} = E_{\text{cl}} + \Delta E + \frac{J(J+1)}{2\alpha^2}, \tag{4.90}$$

one obtains very reasonable results for the nucleon and Δ masses if the constituent quark mass is chosen to be around 400 MeV, see Tab. 4.7. The individual pieces in eq. (4.90) are of the orders $\mathcal{O}(N_c)$, $\mathcal{O}(1)$ and $\mathcal{O}(1/N_c)$, respectively, as expected by general counting rules.

Table 4.7. The predictions for the masses of the nucleon (N) and Δ–resonance. The empirical data are 939MeV and 1232MeV, respectively (taken from ref. [WAR95]).

m(MeV)	$m_\pi = 0$			$m_\pi = 135$MeV		
	400	500	600	400	500	600
E_{cl}(MeV)	1212	1193	1166	1250	1221	1193
ΔE(MeV)	-346	-436	-525	-338	-448	-538
α^2(1/GeV)	6.26	4.73	3.87	5.80	4.17	3.43
M_N(MeV)	926	836	738	976	863	764
M_Δ(MeV)	1166	1153	1126	1236	1223	1201

4.9 Hyperons

Having succeeded to describe the light baryons the question how to include hyperons in the soliton picture of baryons naturally arises. The description of hyperons within the soliton approach depends on whether one considers the strange quark to be light or heavy. In the first case SU(3) is treated as an approximate symmetry calculating symmetry breaking effects in perturbation theory. In the second case one considers strange degrees of freedom as fluctuations off the soliton. In Tab. 4.8 we summarized the differences of the two approaches.

Table 4.8. Comparison of the collective and the bound state approach.

Light strange quark Collective approach	Heavy strange quark Bound state approach
Symmetry breaking small ↓ strangeness as **collective coordinates** (analogous to zero modes) ↓ Collective Hamiltonian Diagonalization ↓ *HYPERONS*	Symmetry breaking large ↓ restoring force for **strange fluctuations** ("harmonic" potential) ↓ Bound State Energy and Wave Function ↓ *HYPERONS*

4.9.1 Collective Approach

First, we will consider the collective approach postponing the bound state approach to the second part of this section. In order to project on baryon quantum numbers we employ the cranking procedure in flavor SU(3) [WAR92],

$$\xi(\boldsymbol{x}, t) = R(t)\xi(\boldsymbol{x})R^{\dagger}(t), \tag{4.91}$$

where $R(t)$ describes the (adiabatic) rotation in SU(3) flavor space. Of course, we only have zero modes in the subspace $SU(2)_I \times U(1)_Y$ of SU(3), nevertheless eq. (4.91) is reasonable since we consider SU(3) as an approximate symmetry. Elevating $R(t)$ to be SU(3) valued furthermore allows us to easily make contact with the phenomenology of baryon representations. This approach has received intensed recognition in Skyrme type models after Yabu and Ando [YA88] demonstrated that the resulting collective Hamiltonian may be diagonalized exactly.

As the cranking of the soliton proceeds somewhat analogous to the SU(2) case we will be very brief here and concentrate on the issues specific to the SU(3) case. The difference of strange quark mass to the light quark masses induces terms not present in the isospin symmeytric two flavor case. So, the contribution to the one–particle Hamiltonian h due to $SU(3)$ symmetry breaking is non–vanishing and given by

$$h_{SB} = T\beta[R^{\dagger}(t)\langle\Sigma\rangle R(t) - \langle\Sigma\rangle]T^{\dagger}$$

$$= \frac{\Delta M}{\sqrt{3}} T \left[\beta \sum_{i=1}^{3} (D_{8i}\lambda_i + \sum_{\alpha=4}^{7} D_{8\alpha}\lambda_\alpha + (D_{88} - 1)\lambda_8)\right]T^{\dagger} \tag{4.92}$$

with $T = \xi(\boldsymbol{x})P_L + \xi^{\dagger}(\boldsymbol{x})P_R$ and $\Delta M = m_u - m_s$ denotes the difference of the up and strange constituent masses. The adjoint representation $D_{ij} = \frac{1}{2}\mathrm{tr}(\lambda_i R\lambda_j R^{\dagger})$ of the rotation matrices clearly exhibits the transformation properties of the symmetry breaking part of the Hamiltonian.

Now, our goal is to expand the energy in both, ΔM and the angular velocities Ω_a. In this expansion we keep terms quadratic in the perturbation $h_P = h_{rot} + h_{SB}$. Then, due to isospin invariance, the most general form of the collective Lagrangian reads

$$L = -E + \frac{1}{2}\alpha^2 \sum_{i=1}^{3} \Omega_i^2 + \frac{1}{2}\beta^2 \sum_{\alpha=4}^{7} \Omega_\alpha^2$$

$$- \frac{\sqrt{3}}{2}B\Omega_8 + \alpha_1 \sum_{i=1}^{3} D_{8i}\Omega_i + \beta_1 \sum_{\alpha=4}^{7} D_{8\alpha}\Omega_\alpha$$

$$- \frac{1}{2}\gamma(1 - D_{88}) - \frac{1}{2}\gamma_1(1 - D_{88}^2) - \frac{1}{2}\gamma_2 \sum_{i=1}^{3} D_{8i}D_{8i} - \frac{1}{2}\gamma_3 \sum_{\alpha=4}^{7} D_{8\alpha}D_{8\alpha}.$$

$$\tag{4.93}$$

E hereby is the classical soliton energy. We will not describe the origin of all these terms in the collective Lagrangian but rather refer the reader to the literature [WAR92].

In order to quantize the collective coordinates Noether charges corresponding to right $SU(3)$ transformations may be constructed leading to the quantization prescription for the right $SU(3)$ generators,

$$R_a = -\frac{\partial L}{\partial \Omega_a} = \begin{cases} -(\alpha^2 \Omega_a + \alpha_1 D_{8a}) = -J_a, & a=1,2,3 \\ -(\beta^2 \Omega_a + \beta_1 D_{8a}), & a=4,..,7 \\ \frac{\sqrt{3}}{2} B, & a=8 \end{cases} \tag{4.94}$$

wherein J_i $(i = 1, 2, 3)$ denote the spin operators.

The Hamiltonian operator

$$H = -\sum_{a=1}^{8} R_a \Omega_a - L \tag{4.95}$$

may be diagonalized exactly which is done along the lines of the original Yabu–Ando approach [YA88] yielding the energy expression for baryon B

$$E_B = E + \frac{1}{2}\left(\frac{1}{\alpha^2} - \frac{1}{\beta^2}\right)J(J+1) - \frac{3}{8\beta^2} + \frac{1}{2\beta^2}\epsilon_{SB}, \tag{4.96}$$

wherein ϵ_{SB} is the eigenvalue of

$$C_2 + \beta^2\gamma(1 - D_{88})$$

$$+ \beta^2(\alpha_1/\alpha^2)\sum_{i=1}^{3} D_{8i}(2R_i + \alpha_1 D_{8i}) + \beta_1 \sum_{\alpha=4}^{7} D_{8\alpha}(2R_\alpha + \beta_1 D_{8\alpha})$$

$$+ \beta^2\gamma_1(1 - D_{88}^2) + \beta^2\gamma_2 \sum_{i=1}^{3} D_{8i}D_{8i} + \beta^2\gamma_3 \sum_{\alpha=4}^{7} D_{8\alpha}D_{8\alpha}. \tag{4.97}$$

$C_2 = \sum_{a=1}^{8} R_a^2$ denotes the quadratic Casimir operator of $SU(3)$. The eigenvalues ϵ_{SB} are determined using a generalized YA approach [PW91, WAR92] (Yabu and Ando only considered $C_2 + \beta^2\gamma(1 - D_{88})$, c.f. ref. [YA88]).

Here we want to point out an important and very interesting consequence of this quantization procedure: The constraint for the right hypercharge $Y_R = \frac{2}{\sqrt{3}}R_8 = B$ confines the possible eigenstates to those which carry half integer spin for $B = 1$, i.e. fermions. Therefore the fermionic nature of baryons (made non–linearly of mesons, i.e. bosons) comes out naturally.

In Tab. 4.9 the resulting predictions for the hyperon mass spectrum are displayed. As the quantum corrections of $O(N_c^0)$ are not known in the three flavor case we give mass differences only. As may been seen from Tab. 4.9 the mass splittings between states of different spins decrease with increasing constituent mass. This is mainly linked to the decrease of α^2 with m_u. On the contrary the mass differences between members of the same spin multiplet get smaller as the constituent mass gets larger since also the coefficient of the most important symmetry breaking term, γ, decreases with m_u. We furthermore recognize from Tab. 4.9 that the mass splittings between baryons of the same spin are predicted too low. This shortcoming is not completely unexpected since already in the

Table 4.9. The mass differences for the low-lying $\frac{1}{2}^+$ and $\frac{3}{2}^+$ baryon states as functions of the up-constituent mass m_u. All numbers are in MeV. (Taken from ref. [WAR92].)

m_u	350	400	450	500	Exp.
M_N	1684	1725	1726	1717	938
$M_\Lambda - M_N$	149	116	94	78	177
$M_\Sigma - M_N$	182	157	134	116	254
$M_\Xi - M_N$	320	256	210	177	379
$M_\Delta - M_N$	209	296	350	392	293
$M_{\Sigma^*} - M_N$	350	401	433	460	446
$M_{\Xi^*} - M_N$	493	506	515	526	591
$M_\Omega - M_N$	637	611	596	592	733

meson sector of the model we have found that $e.g.$ for $m_u = 400\text{MeV}$ f_K is about 15% lower than the experimental value. That this deficiency transfers to the baryon sector may be seen from the following estimate. We choose $m_u = 390MeV$ to reproduce the experimental nucleon-Δ mass difference. Then we scale the dominant symmetry breaking parameter $\gamma^m + \gamma^{\text{vac}}$ by $(\frac{f_K^{expt.}}{f_K^{pred.}})^2 = (\frac{114}{100})^2$ to calculate the baryon mass differences. The valence quark contribution to γ should be described only by the constituent masses but not by the predicted value for f_K. The results stemming from this estimate are shown in Tab. 4.10 and are found to be in excellent agreement with experimental data.

Table 4.10. The mass differences for the estimate described in the text. The particles' names refer to the corresponding mass difference with respect to the nucleon. All numbers are in MeV. (Taken from ref. [WAR92].)

$m_u = 390\text{MeV}$	Λ	Σ	Ξ	Δ	Σ^*	Ξ^*	Ω
calculated	121	162	267	283	394	504	615
γ_{rescaled}	175	248	396	291	449	608	765
Exp.	177	254	379	293	446	591	733

Thus we may summarize the results for the collective approach by stating that the too small predictions for the baryon mass splittings in the collective approach are directly connected to the too small result for f_K. We furthermore expect that a more complete model ($i.e.$ (axial-) vector mesons included) which reproduces the experimental value for f_K will in fact yield agreement for the baryon mass differences.

4.9.2 Bound State Approach

Having now calculated the baryon mass splittings in the collective approach we will treat the hyperons in the bound state approach. First, we will evaluate the energy eigenvalue of the kaon bound by the soliton in the presence of SU(3) symmetry breaking. This treatment was originally set forth by Callan and Klebanov for the Skyrme model to describe strange baryons [CK85]. Based on the formalism for fluctuations off the soliton we consider meson fluctuations into strange direction only

$$\eta(\mathbf{r}, t) = 0 \quad \text{and} \quad \sum_{\alpha=4}^{7} \eta_\alpha(\mathbf{r}, t)\lambda_\alpha = \begin{pmatrix} 0 & K(\mathbf{r}, t) \\ K^\dagger(\mathbf{r}, t) & 0 \end{pmatrix} \tag{4.98}$$

wherein $K(\mathbf{r}, t)$ is a two-component isospinor. The corresponding Fourier transform reads

$$\sum_{\alpha=4}^{7} \tilde{\eta}_\alpha(\mathbf{r}, \omega)\lambda_\alpha = \begin{pmatrix} 0 & \tilde{K}(\mathbf{r}, \omega) \\ \tilde{K}^\dagger(\mathbf{r}, -\omega) & 0 \end{pmatrix}. \tag{4.99}$$

In the case of SU(3) symmetry ($m_s^0 = m^0$) the associated zero mode is given by

$$K_0(\mathbf{r}) = \hat{\mathbf{r}} \cdot \boldsymbol{\tau} U_0 \begin{pmatrix} \sin\frac{\Theta(r)}{2} \\ 0 \end{pmatrix} \tag{4.100}$$

where U_0 is an arbitrary 2×2 space-time independent unitary matrix fixing the isospin orientation. Eq. (4.100) is obtained analogously to the isorotational zero modes, see Sect. 4.7. In this case one parametrizes the infinitesimal vector transformation of the chiral soliton (4.59) into strange direction in terms of the fluctuation (4.98). $K_0(\mathbf{r})$ obviously carries isospin $T = 1/2$, orbital angular momentum $L = 1$ and grand spin $G = 1/2$. A kaon bound state develops in this channel once SU(3) symmetry is abandoned and physical parameters are assumed. We thus employ the ansatz

$$\tilde{K}(\mathbf{r}, \omega) = \hat{\mathbf{r}} \cdot \boldsymbol{\tau} \Omega(r, \omega) \tag{4.101}$$

for the kaon bound state. $\Omega(r, \omega)$ is a two component isospinor which only depends on the radial coordinate r and the frequency ω, i.e. the angular dependence is separated. Using eq. (4.101) the perturbative parts of the hamiltonian are relatively simple,

$$\tilde{h}_{(1)}(\mathbf{r}, \omega) = -\frac{1}{2}(m_u + m_s) \begin{pmatrix} 0 & u_0(r)\Omega(r, \omega) \\ \Omega^\dagger(r, -\omega)u_0(r) & 0 \end{pmatrix} \tag{4.102}$$

$$\tilde{h}_{(2)}(\mathbf{r}, \omega, -\omega) = \frac{1}{4}(m_u + m_s)$$
$$\begin{pmatrix} u_0(r)\beta\Omega(r, \omega)\Omega^\dagger(r, \omega)u_0(r) & 0 \\ 0 & -\beta\Omega^\dagger(r, -\omega)\Omega(r, -\omega) \end{pmatrix} \tag{4.103}$$

For convenience we have introduced the unitary, self-adjoint transformation matrix u_0

$$u_0(\mathbf{r}) = \beta \left(\sin\frac{\Theta}{2} - i\gamma_5 \hat{\mathbf{r}} \cdot \boldsymbol{\tau}\cos\frac{\Theta}{2} \right). \qquad (4.104)$$

The main task now consists of computing matrix elements of the form

$$\left| _{\text{ns}}\langle\mu|\tilde{h}_{(1)}|\nu\rangle_{\text{s}} \right|^2, \qquad _{\text{ns}}\langle\mu|\tilde{h}_{(2)}|\nu\rangle_{\text{ns}} \qquad \text{and} \qquad _{\text{s}}\langle\mu|\tilde{h}_{(2)}|\nu\rangle_{\text{s}} \qquad (4.105)$$

with $|\mu\rangle_{\text{ns}} = \sum_\rho V_{\rho\mu}|\rho\rangle_{\text{ns}}^0$ denoting SU(2) hedgehog states while $|\nu\rangle_{\text{s}}$ refers to the (unperturbed) states of the strange quark. The orthogonal matrix V is obtained by diagonalizing $h_{(0)}$ in the non-strange sector and solving the equation of motion for the hedgehog (C.6) self-consistently. The evaluation of the matrix elements (4.105) is straightforward but lengthy, however, the resulting expression for the action may again be written with the help of kernels as

$$\mathcal{A}[\Omega] = \int_{-\infty}^{+\infty} \frac{d\omega}{2\pi} \Bigg\{ \int dr\, r^2 \int dr'\, r'^2 \; \Phi^{(2)}(\omega; r, r')\Omega^\dagger(r, \omega)\Omega(r', \omega) $$
$$+ \int dr\, r^2 \; \Phi^{(1)}(r)\Omega^\dagger(r, \omega)\Omega(r, \omega) \Bigg\} \qquad (4.106)$$

For the explicit form of these kernels we refer to [WAR94]. The equation of motion for $\Omega(r, \omega)$ is then the Bethe–Salpeter equation

$$r^2 \left\{ \int dr'\, r'^2 \Phi^{(2)}(\omega; r, r')\Omega(r', \omega) + \Phi^{(1)}(\omega; r)\Omega(r, \omega) \right\} = 0. \qquad (4.107)$$

This Bethe-Salpeter equation in the soliton background corresponds to the bound state equation of the Callan-Klebanov approach [CK85] to the Skyrme model.[¶]
$K(\mathbf{r}, t)$ represents an $s\bar{q}$ ($\bar{s}q$) excitation which, according to eq. (4.79), converts a non-strange valence quark into a strange valence (anti-) quark. The emerging strange valence (anti-) quark carries total angular momentum $J = 1/2$ and isospin $I = 0$. The spinor representation reads

$$\delta\Psi_{\text{val}}^s(\mathbf{r}, t) = \int_{-\infty}^{+\infty} \frac{d\omega}{2\pi} \delta\tilde{\Psi}_{\text{val}}^s(\mathbf{r}, \omega)\, e^{-i(\epsilon_{\text{val}} - \omega)t} \qquad (4.108)$$

$$\delta\tilde{\Psi}_{\text{val}}^s(\mathbf{r}, \omega) = -\frac{1}{2}(\hat{m} + m_s)\left(\epsilon_{\text{val}} - \omega - h_{(0)}(\mathbf{r})\right)^{-1}$$
$$\times \Omega^\dagger(r, \omega)\left(\sin\frac{\Theta}{2} - i\hat{\mathbf{r}} \cdot \boldsymbol{\tau}\gamma_5\cos\frac{\Theta}{2}\right)\Psi_{\text{val}}^{\text{ns}}(\mathbf{r}) \qquad (4.109)$$

wherein we have separated the time dependence of the non-strange valence quark. The interpretation of eqs. (4.108) and (4.109) is obvious: A kaon fluctuation with frequency ω changes a non–strange valence quark of energy ϵ_{val}

[¶] As usual the normalization of the bound state wave function remains undetermined by the Bethe–Salpeter equation. For a consistent normalization we need to calculate the strangeness of the kaon fluctuations using the same arbitrary units as in the Bethe–Salpeter equation, see ref. [WAR94].

into a strange valence (anti–) quark with the energy $\epsilon_{\text{val}}^{\text{s}} = \epsilon_{\text{val}} - \omega$. For $\epsilon_{\text{val}}^{\text{s}} > 0$ this state has to be identified with a strange quark carrying strangeness $S = -1$ while for $\epsilon_{\text{val}}^{\text{s}} < 0$ an $S = +1$ anti-quark is induced. For $\epsilon_{\text{val}} > 0$ a baryon with vanishing strangeness is constructed by occupying the valence quark state N_c-times. The contribution of the valence quarks to the energy of this $(S = 0)$ baryon is therefore $N_c \epsilon_{\text{val}}$. The occupation of a kaon mode with frequency ω replaces this non-strange baryon by a baryon with strangeness -1 (+1) for $\epsilon_{\text{val}}^{\text{s}} > 0$ ($\epsilon_{\text{val}}^{\text{s}} < 0$). The corresponding contribution of the valence quarks to the energy of this hyperon hence is $N_c \epsilon_{\text{val}} - \omega$. This discussion reveals that a <u>negative</u> bound state energy ω is required in order to describe physical hyperons.

The system consisting of the chiral soliton and the kaon bound state - as it stands - does not describe physical baryons although it carries unit baryon number. It actually corresponds to a superposition of states with arbitrary spin and isospin. In order to project onto states with good spin and isospin we apply again the semiclassical quantization

$$M = R(t)\xi_0 \xi_f \langle \Sigma \rangle \xi_f \xi_0 R^\dagger(t) \qquad \text{with} \qquad R(t) \in SU(2). \qquad (4.110)$$

Owing to the isosinglet character of $\langle \Sigma \rangle$ the *ansatz* (4.110) is equivalent to the substitution

$$\xi_0 \rightarrow R(t)\xi_0 R^\dagger(t) \qquad \text{and} \qquad \tilde{K} \rightarrow R(t)\tilde{K}. \qquad (4.111)$$

This identification especially implies that the total isospin is carried by the collective coordinates $R(t)$ while the kaon field $\tilde{K}$ has lost its isospin. As the NJL model is originally formulated in terms of quark fields this is intuitively clear since this "transmutation" of isospin is a direct consequence of the fact that the strange quarks (4.109) have zero isospin. Moreover, eq. (4.111) shows that due to the hedgehog structure of ξ_0 the absolute value of the spin carried by the soliton is identical to the total isospin.

Now we want to expand the action in powers of Ω in the presence of the kaon bound state. The term quadratic in Ω yields then again the moment of inertia α^2 for isorotations, see eq. (4.48). Our special attention, however, concerns the coupling of the kaon fluctuations to the angular velocities which will determine the mass splittings of baryons with different isospin but equal strangeness. The lowest order term which contains this coupling and represents an isosinglet is linear in Ω and quadratic in the kaon fluctuations. Assuming that there is exactly one kaon bound state (which results from the numerical calculation, see below) we will consider only those baryon states which are constructed by the soliton and this state. Introducing additionally annihilation and creation operators for this state the collective Lagrangian due to the rotations reads

$$L_\Omega = \frac{1}{2}\alpha^2 \Omega^2 - \frac{1}{2}c\, \Omega \cdot \left(\sum_{i,j=1}^{2} a_i^\dagger \tau_{ij} a_j \right) + \dots \qquad (4.112)$$

where the quantity c is given by mode sums over quark eigenstates [WAR94]. Owing to eq. (4.111) and the hedgehog structure of the soliton the momentum conjugate to Ω is the spin carried by the soliton

$$\mathbf{J}_\Theta = \frac{\partial L_\Omega}{\partial \Omega} = \alpha^2 \Omega - \frac{1}{2} c\Omega \cdot \left(\sum_{i,j=1}^{2} a_i^\dagger \tau_{ij} a_j \right) + \ldots \qquad (4.113)$$

In order to extract the spin carried by the strange fluctuations we next consider the total spin which by definition is the expectation value

$$\langle \mathbf{J} \rangle = \int D\bar{q} Dq \int d^3r \; q^\dagger \mathbf{J} q \, \exp\left(iA_{\mathrm{NJL}} \right) \qquad (4.114)$$

where A_{NJL} denotes the action associated with the NJL Lagragian $\mathcal{L}_{\mathrm{NJL}}$ (3.1). Since the spin operator commutes with the collective rotations $R(t)$ we can straightforwardly transform to the rotating frame

$$\langle \mathbf{J} \rangle = \int D\bar{q}' Dq' \int d^3r \; q'^\dagger \mathbf{J} q' \, \exp\left(iA'_{\mathrm{NJL}} \right). \qquad (4.115)$$

Here A'_{NJL} represents the NJL action in the rotating frame which also contains the Coriolis term

$$A'_{\mathrm{NJL}} = \int d^4x \left(\mathcal{L}_{\mathrm{NJL}} - \frac{1}{2} q'^\dagger \tau \cdot \Omega q' \right). \qquad (4.116)$$

According to the definition of the grand spin $\mathbf{G}$ we may rewrite eq. (4.115) as

$$\langle \mathbf{J} \rangle = \int D\bar{q}' Dq' \int d^3r \; q'^\dagger \left(\mathbf{G} - \frac{\tau}{2} \right) q' \, \exp\left(iA'_{\mathrm{NJL}} \right). \qquad (4.117)$$

In this expression we may identify the soliton contribution to the spin $\mathbf{J}_\Theta$ by differentiating with respect to the angular velocity Ω

$$\langle \mathbf{J} \rangle = \langle \mathbf{G} \rangle + \int D\bar{q}' Dq' \; \frac{1}{T} \frac{\partial A'_{\mathrm{NJL}}}{\partial \Omega} \exp\left(iA'_{\mathrm{NJL}} \right) = \langle \mathbf{G} \rangle + \mathbf{J}_\Theta. \qquad (4.118)$$

The spin carried by the strange fluctuations is just the difference of the total and the soliton spins

$$\mathbf{J}_K = \langle \mathbf{J} \rangle - \mathbf{J}_\Theta = \langle \mathbf{G} \rangle. \qquad (4.119)$$

Thus we have obtained the somewhat surprising result that the kaon spin is identical to the expectation value of the grand spin.

The collective Hamiltonian obtained from L_Ω reads

$$H_\Omega = \frac{1}{2\alpha^2} \left(\mathbf{J}_\Theta + \frac{c}{2} \sum_{i,j=1}^{2} a_i^\dagger \tau_{ij} a_j \right)^2 = \frac{1}{2\alpha^2} \left(\mathbf{J}_\Theta + \chi \mathbf{J}_K \right)^2 \qquad (4.120)$$

where we have introduced the parameter $\chi = -c/d$. Since the total spin is the sum $\mathbf{J} = \mathbf{J}_\Theta + \mathbf{J}_K$ and the absolute value of $\mathbf{J}_\Theta$ equals the total isospin H_Ω becomes

$$H_\Omega = \frac{1}{2\alpha^2}\left(\chi J(J+1) + (1-\chi)I(I+1) + (\frac{d\chi}{2})^2 S(S-2)\right). \quad (4.121)$$

Here J and I denote the spin and isospin of the baryon under consideration. We have furthermore used that $\left(\sum_{i,j=1}^2 a_i^\dagger \tau_{ij} a_j\right)^2$ may be written as $S(S-2)$. This term is already of fourth order in the kaon fluctuations and should thus been dropped for consistency. In order to finally reach the mass formula we note that each occupation of the kaon bound state corresponds to replacing a non–strange valence quark by a strange valence quark. This shifts the total energy by $S\omega_0$ since we have $S \leq 0$ for physical baryons. We may now conclude this section by presenting the mass formula for strange and non–strange baryons

$$M_B = E_{\text{cl}} + S\omega_0 + \frac{1}{2\alpha^2}\left(\chi J(J+1) + (1-\chi)I(I+1)\right) \quad (4.122)$$

wherein E_{cl} is the classical energy due to the static soliton configuration. We would finally like to note that the baryon mass formula (4.122) actually does not depend on the normalization of the strange bound state since $\chi = -c/d$ is the ratio of two objects quadratic in the bound state wave–function.

The bound state energy ω_0 lies between -200 MeV and -150 MeV for reasonable values of the constituent quark masses whereas the parameter χ describing the coupling of bound state and collective SU(2) rotations is approximately 0.40. We have already seen that the moment of inertia α^2 comes out reasonable for a constituent quark mass around 400 MeV. Upon inversion of the mass formula (4.122) we can, on the other hand, get some information of the "empirical" values of the parameters ω_0, α^2 and χ. Since the nucleon and Ξ baryons have both equal spin and isospin we may extract from (4.122)

$$M_\Xi - M_n = -2\omega_0 \approx 379\text{MeV} \quad \text{or} \quad \omega_0 \approx -189.5\text{MeV} \quad (4.123)$$

which also favors a constituent quark mass m_u somewhat below 400MeV. The "empirical" value for the moment of inertia α^2 is given by

$$M_\Delta - M_n = \frac{3}{2\alpha^2} \approx 293\text{MeV} \quad \text{or} \quad \alpha^2 \approx 5.12\text{GeV}^{-1}. \quad (4.124)$$

Finally we get for χ from states with identical strangeness

$$\frac{M_{\Sigma^*} - M_\Sigma}{M_\Sigma - M_\Lambda} = \frac{3\chi}{2(1-\chi)} \approx 2.49 \quad \text{or} \quad \chi \approx 0.62. \quad (4.125)$$

We note that the NJL model underestimates this value somewhat.

In order to judge this approach we compare the results on the baryon spectrum with the one obtained in the collective approach as well as the experimental data. For getting consistent treatments we fix the constituent quark mass in both approaches by demanding the experimental value for nucleon–Δ mass difference

Table 4.11. The mass differences of the low-lying $\frac{1}{2}^{+}$ and $\frac{3}{2}^{+}$ baryons with respect to the nucleon. We compare the predictions of the Callan–Klebanov (CK) and Yabu–Ando (YA) approaches to the NJL model to the experimental data. All numbers are in MeV. (Taken from ref. [WAR94].)

	CK	YA	Expt.
m_u	430	407	—
Λ	132	105	177
Σ	234	148	254
Ξ	341	236	379
Δ	293	293	293
Σ^*	374	387	446
Ξ^*	481	482	591
Ω	613	576	733

$M_\Delta - M_n = 293\,\text{MeV}$ leading to $m_u = 430$ MeV. For the parameters involving the bound state wave function we find in this case $\omega_0 = -170.6\,\text{MeV}, c = -0.42, d = 0.89$ and $\chi = 0.48$. In Tab. 4.11 the results on the baryon mass differences are displayed. Again we find too low predicted mass differences. Probably they are also merely due to the incorrect prediction of the ratio f_K/f_π in the meson sector of the NJL model.

Summary

The Skyrme model (4.1) possesses solitonic solutions. The soliton with winding number one is interpreted as baryon. The emergence of the soliton is understood by considering the quark energy spectrum in the presence of the soliton field: For a sufficiently strong topological meson field with winding number n n quark levels join the negative Dirac sea. Thus we have demonstrated that the identification of the topological winding number with the baryon number (Witten's conjecture) arises naturally within QFD if the solitonic meson fields are strong enough to pull positive–energy levels into the Dirac sea.

The energy functional of QFD for static meson fields has been derived. One obtains the corresponding static solution by minimizing it numerically. If all meson fields are included the valence quark energy is negative supporting thereby the soliton picture of baryons. In order to project on good spin and isospin a semiclassical quantization (cranking) has been performed. This leads to the experimental value for the nucleon–Δ mass splitting for a constituent quark mass slightly above 400 MeV. Meson fluctuations off the soliton have been considered. Especially, the zero modes due to broken rotational and translational invariance have been identified. Including their contribution to quantum corrections the absolute values of the nucleon and Δ masses are well reproduced. Finally, hyperons have been described within the two different approaches. In the collective approach strangeness is treated as collective coordinate and the resulting Hamiltonian is diagonalized. In the bound state approach hyperons are considered as being bound states of kaons and solitons, *i.e.* in the quark language a (light) valence quark is substituted by a strange quark.

Chapter 5

Baryons as Bound States of Diquarks and Quarks

5.1 Functional Integral Hadronization

In the chapters three and four we have completely neglected the quark–quark (as compared to the antiquark–quark) interaction of QFD, *i.e.* only the first term in eq. (2.55) has been taken into account. This was justified for large N_c which is the conceptual basis of the soliton picture. Now we will consider $N_c = 3$ and investigate the second term of the Lagrangian (2.55). We slightly generalize it by allowing different coupling constants g_1 and g_2 in the meson and diquark channels, respectively

$$\mathcal{L}_{int} = \frac{g_1}{3}(\bar{q}\Lambda_\alpha q)(\bar{q}\Lambda^\alpha q) + \frac{g_2}{3}(\bar{q}\Sigma_\alpha q^C)(\bar{q}^C \Sigma^\alpha q). \tag{5.1}$$

Here q^C and $\bar{q}^C$ are the charge conjugated spinors. For their definition as well as the one of the vertices Λ_α and Σ_α see Sect. 2.3, especially eqs. (2.56,2.57). Note that $\mathcal{L}_{int}^{\bar{q}q}$ and $\mathcal{L}_{int}^{qq}$ are both separately invariant under chiral transformations. From the phenomenological point of view independent coupling constants g_1 and g_2 in the two channels are therefore tolerated by chiral symmetry.

The aim is to transform the effective quark theory defined by (5.1) into an effective theory of the physical mesons and baryons where for a finite number of colors, $N_c = 3$, the latter are considered as quark–diquark bound states. In the approach [Re90] which we will follow here one introduces composite meson and baryon fields built up from the quark degrees of freedom in accord with their valence quark content. Taking into account an explicit baryon field is the main difference as compared to the functional integral bosonization of the antiquark–quark interaction only, which was performed in Sect. 3.1. We will therefore concentrate on the part involving the baryon field. It may be introduced via the identity

$$
1 = \int \prod_{\alpha,\mu} \mathcal{D}\bar{\Psi}_\mu^\alpha \mathcal{D}\Psi_\mu^\alpha \ \delta(\bar{\Psi}_\mu^\alpha - \bar{q}_\mu(\bar{q}\Sigma^\alpha q^C))\delta(\Psi_\mu^\alpha - (\bar{q}^C\Sigma^\alpha q)q_\mu)
$$

$$
= \int \prod_{\alpha,\mu} \mathcal{D}\bar{\Psi}_\mu^\alpha \mathcal{D}\Psi_\mu^\alpha \mathcal{D}\bar{\chi}_\mu^\alpha \mathcal{D}\chi_\mu^\alpha
$$

$$
\exp\Big\{ i \int d^4x \ d^4X \Big((\bar{\Psi}_\mu^\alpha(x,X) - \bar{q}_\mu(x)(\bar{q}(X)\Sigma^\alpha q^C(x)))\chi_\mu^\alpha(x,X)
$$

$$
+ \bar{\chi}_\mu^\alpha(x,X)(\Psi_\mu^\alpha(x,X) - (\bar{q}^C(X)\Sigma^\alpha q(X))q_\mu(x)) \Big) \Big\}
$$

$$(5.2)$$

where the colorless three quark configuration has been expressed by a diquark in the $\bar{3}_c$ representation $\bar{q}^c\Sigma^\alpha q$ and a quark q_μ. The color index a in $\bar{q}^c\Sigma^\alpha q$ (see eq. (2.57)) is now a dummy index summed over with the corresponding index in q_μ. The remaining explicit index α in the diquark combination then denotes collectively only flavor and Dirac indices. In view of the importance of diquarks as building blocks of baryons and also because of the presence of the diquark interaction $\mathcal{L}_{int}^{qq}$ it is convenient to introduce explicitly diquark fields as well:

$$
1 = \int \prod_\alpha \mathcal{D}\kappa_\alpha \mathcal{D}\kappa_\alpha^* \delta(\kappa_\alpha^* - \bar{q}\Sigma_\alpha q^C)\delta(\kappa_\alpha - \bar{q}^C \Sigma_\alpha q)
$$

$$
= \int \prod_\alpha \mathcal{D}\kappa_\alpha \mathcal{D}\kappa_\alpha^* \mathcal{D}\Delta_\alpha \mathcal{D}\Delta_\alpha^* \exp\Big\{ i \int d^4x \Big[(\kappa_\alpha^*(x) - \bar{q}(x)\Sigma_\alpha q^C(x))\Delta^\alpha(x)
$$

$$
+ \Delta_\alpha^*(x)(\kappa^\alpha(x) - \bar{q}^C(x)\Sigma^\alpha q(x)) \Big] \Big\}.
$$

$$(5.3)$$

By constructions κ and Δ are complex bose fields while Ψ and χ are fermion fields. From the symmetry relations

$$
(\bar{q}\Sigma_\alpha q^C)^\dagger = \pm(\bar{q}\Sigma_\alpha q^C)
$$

$$(5.4)$$

follows that (depending on α) $\kappa_\alpha, \Delta_\alpha$ and $\kappa_\alpha^*, \Delta_\alpha^*$ are either complex conjugate or anti–complex conjugate to each other. Inserting the identities (5.2) and (5.3) into the generating functional and exploiting the constraints defined by the δ-functions one can replace $\bar{q}^c\Sigma_\alpha q$ and $\bar{q}\Sigma_\alpha q^c$ by κ_α and κ_α^*, respectively. The bosonization of meson fields can be done as described in Sect. 3.1. As a result the quark fields are removed from the two–body interactions and can be integrated out exactly. Furthermore, the fields κ, κ^* can also be integrated out. The quantum transition amplitude, $i.e.$ the generalized form of the quantity (3.5) containing explicit baryon fields, can then be cast into the form

$$
Z_{\text{QFD}} = \int \mathcal{D}\eta \mathcal{D}\bar{\Psi}\mathcal{D}\Psi\mathcal{D}\bar{\chi}\mathcal{D}\chi \ \exp\Big(iW[\eta,\chi] + i\int(\bar{\Psi}\chi + \bar{\chi}\Psi)\Big)
$$

$$(5.5)$$

where $W[\eta, \chi]$ is the generating functional of the baryon (three–quark) Green's function in the background of the fluctuating meson field η with χ figuring as colorless three–quark auxiliary field which, in principle, should be integrated out. The generating functional $W[\eta, \chi]$ is still exact but unfortunately only implicitly defined as functional integral over the diquark field Δ

$$W[\eta, \chi] = -i \log \int \mathcal{D}\Delta \mathcal{D}\Delta^* \exp\left(i\mathcal{A}_{\mathrm{eff}}[\eta, \Delta, \chi]\right) \tag{5.6}$$

where the effective action $\mathcal{A}_{\mathrm{eff}}$ is defined by

$$\mathcal{A}_{\mathrm{eff}}[\eta, \Delta, \chi] = -3 \int d^4x \left(\frac{\eta^2}{4g_1} + \frac{\Delta^*\Delta}{4g_2}\right) - \frac{i}{2}\mathrm{Tr}\log\mathcal{G}^{-1}[\eta, \Delta, \chi]$$
$$-\frac{1}{2}\left(\frac{3}{2g_2}\right)^2 \int d^4x \left((\bar{\chi}\Delta) \quad -(\Delta^*\chi)^T\right) \mathcal{G}\left(\begin{array}{c}(\Delta^*\chi)\\-(\bar{\chi}\Delta)^T\end{array}\right). \tag{5.7}$$

The Green's function of the quarks in the fluctuating meson field η and in the auxiliary baryon field $\chi(x, X)$, $\mathcal{G}[\eta, \Delta, \chi]$, is defined via

$$\mathcal{G}^{-1}[\eta, \Delta, \chi](x, y) = \begin{pmatrix} G^{-1}[\eta, \chi](x, y) & -\Delta C\delta(x - y) \\ -C\tilde{\Delta}\delta(x - y) & (G^{-1}[\eta, \chi])^T(x, y) \end{pmatrix} \tag{5.8}$$

and

$$G^{-1}[\eta, \chi](x, y) = i\not{D}\,\delta(x - y) - R(x, y)$$
$$i\not{D} = i\not{\partial} - m^0 - \eta$$
$$R(x, y) = \frac{3}{g_2} \int d^4z\, \chi^\alpha(x, z)\bar{\chi}^\alpha(y, z). \tag{5.9}$$

In the above expressions all suppressed indices are summed over, and we have introduced the meson and diquark matrix fields

$$\eta = \eta_\alpha \Lambda^\alpha, \quad \Delta = \Delta_\alpha \Sigma^\alpha, \quad \tilde{\Delta} = \Delta_\alpha^* \Sigma^\alpha. \tag{5.10}$$

The matrix (5.8) is the Green's function in the Nambu–Gorkov formalism [Go59, Na60] of BCS theory, see e.g. Chap. 13 of ref. [FW71]. It contains besides the usual Green's function (5.9) which is the time–ordered expectation value of the quark–antiquark operator,

$$G[\eta, \chi](x, y) = -i\langle T(q(x)\bar{q}(y))\rangle, \tag{5.11}$$

also the so–called anomalous Green's function which gives the probability for the creation or annihilation of a diquark out of the correlated ground state,

$$\langle T(q(x)\bar{q}^c(y))\rangle, \quad \langle T(q^c(x)\bar{q}(y))\rangle, \tag{5.12}$$

which are related to $\tilde{\Delta}$ and Δ, respectively. Note that the corresponding condensates

$$\langle q(x)\bar{q}^c(x)\rangle, \quad \langle q^c(x)\bar{q}(x)\rangle \tag{5.13}$$

vanish as long as color symmetry is not spontaneously broken.

Since Δ and χ are not observable fields one would like to integrate them out leaving a theory defined entirely in terms of the physically observable meson and baryon fields. However, this can be accomplished only approximately by resorting to an expansion of the generating functional in terms of the field χ

$$W[\eta, \chi] = \mathcal{A}[\eta] + \int d^4x \, d^4y \, d^4X \, d^4Y \; \bar{\chi}(x, X)G_B[\eta](x, X, y, Y)\chi(y, Y) + \dots .$$
(5.14)

Here $\mathcal{A}[\eta] = W[\eta, \chi = 0]$ is the effective mesonic action (3.11)* and

$$G_B[\eta](x, X, y, Y) = \left(\frac{\delta^2 W[\eta, \chi]}{\delta\bar{\chi}(x, X)\delta\chi(y, Y)} \right)_{\chi=0}$$
(5.15)

is the connected baryon Green's function after projection onto physical flavor channels. The omitted higher order terms represent baryon correlations that cannot be generated by meson exchange, which, however, can be modified or dressed by mesonic exchanges. Such correlations are also obtained in traditional quark models. Since we have at present little empirical evidence for these correlations we will ignore them in the following.

Truncating the expansion (5.14) at second order the χ field can be integrated out in Gaussian approximation:

$$Z_{\text{QFD}}[\bar{\xi}, \xi] = \int \mathcal{D}\eta \exp(i\mathcal{A}[\eta])\text{Det}G_B[\eta]$$

$$\int \mathcal{D}\bar{\Psi}\mathcal{D}\Psi \exp\left(i\int \bar{\Psi}G_B^{-1}[\eta]\Psi + \bar{\xi}\Psi + \bar{\Psi}\xi \right).$$
(5.16)

We have included explicit baryon sources ξ and $\bar{\xi}$. This generating functional describes a non–local coupled meson and baryon quantum field theory where the baryons move in the background of the fluctuating meson field η. This is indicated by writing the baryon propagator as a functional of meson fields, $G_B[\eta]$. The baryon determinant $\text{Det}G_B[\eta]$ may, at first sight, appear strange, since phenomenological quantum hadron field theories (*e.g.* the Walecka model [SW85]) do not have it. However, this term, which naturally arises in the above hadronization procedure, is quite important: It makes the baryonic part of the effective hadron theory (5.16) anomaly free. The Jacobi determinant related to an anomaly, in particular the chiral anomaly (see Sect. 3.7), which arises from the non–invariance of the measure of the functional integral over the baryon fields, is canceled by the baryon determinant. Thus double counting of anomalous (meson) processes with baryon loops is avoided. The anomalies arise also for the effective action (5.16) from the fundamental fermions, the quarks, and are entirely included already in the effective mesonic action $\mathcal{A}[\eta]$ (3.11) through the quark determinant as discussed in Sect. 3.7.

---

* Note that we did not shift the meson field, $\eta \to \eta - m_0$, which can be, however, done in a straightforward manner.

If one is not explicitly interested in baryons one may set the sources $\bar{\xi}$ and ξ equal to zero leaving the effective meson theory (3.11) as the integration over baryons cancel exactly the baryon determinant. In fact, as we have seen in Sect. 2.3, in the limit of a large number of colors $N_c \rightarrow \infty$ the attractive diquark interaction $\mathcal{L}_{int}^{qq}$ in (5.1) is suppressed as $g_2/N_c \rightarrow 0$. In this limit there is no need to introduce diquark or explicit baryon fields, *i.e.* baryons are then described as solitons of mesonic fields. The introduction of the meson field η is then sufficient to eliminate the unobservable quark fields. The effective meson theory (3.11) follows also formally from eqs. (5.16) and (5.7) in the limit $g_2/N_c \rightarrow 0$, which constrains the diquark field to $\Delta = 0$ in order to keep the action finite. This shows that the hadronization scheme presented here is consistent with the bosonization described in Sect. 3.1. The obtained reduction of the effective quark theory to the effective meson theory for $N_c \rightarrow \infty$ is consistent with the general observation by 't Hooft and Witten that for a large number of colors QCD reduces to an effective theory of mesons and glueballs [tH74, Wi79]. For $N_c = 3$ the diquark correlations are included and the effective meson action is renormalized. It is no longer given by the "simple" expression (3.11) but implicitly by an integral over the diquark fields $W[\eta, \chi = 0]$, see eq. (5.14). The physical baryons are therefore expected to carry both features: *In the physical baryons the two complementary pictures, chiral solitons and diquark-quark bound state, are competing.*

5.2 Diquarks

Having discussed the soliton nature of baryons in Chap. 4 in great detail we will now focus on the picture of baryons as three–quark bound states. We have seen that diquarks appear as important building blocks in this approach. Therefore we will discuss the properties of diquarks in this section.

First, we want to discuss the possible Dirac and flavor quantum numbers of diquarks. We have already mentioned in Sect. 2.3 that these are restricted by the Pauli principle. Only color antitriplet diquarks are allowed in constructing physical colorless baryons. Thus the qq vertex $C\Sigma^\alpha$ is required to be antisymmetric in its color indices. For three flavors the Pauli principle then enforces that for diquarks in the $\bar{3}_F$ and 6_F flavor representation the spinor matrix is an element of $\{1, i\gamma_5, i\gamma^\mu\gamma^5/\sqrt{2}\}$ or $\{i\gamma^\mu/\sqrt{2}\}$, respectively. Therefore only the axialvector diquark is symmetric under flavor exchange whereas all other diquarks are antisymmetric in flavor. Note that the intrinsic parity of a fermion–fermion pair is exactly opposite to the intrinsic parity of a antifermion–fermion pair. Therefore the combination of two quark spinors, $\bar{q}^C q$ and $\bar{q}q^C$, with *e.g.* $i\gamma_5$ is a scalar and that with γ^μ is an axialvector. Note also that for two flavors there is only one flavor combination (proportional to τ_2) for the scalar diquark. In the following we will concentrate especially on this diquark.

We will now extract the diquark Lagrangian from the effective action (5.7). For this purpose it is sufficient to consider the case of a vanishing three-quark field, $\chi = 0$, and to fix the meson field at its vacuum expectation value, $\eta = \langle \Phi \rangle$. As discussed in Chap. 3 $\langle \Phi \rangle$ is a scalar and diagonal in flavor, the diagonal matrix elements being the quark constituent masses m_i, $i = u, d, s$. This simplified effective action reads then

$$\mathcal{A}_\Delta[\Delta] = -\frac{3}{4g_2} \int d^4x \, \Delta_\alpha^* \Delta_\alpha - \frac{i}{2} \mathrm{Tr} \log \mathcal{G}^{-1} \qquad (5.17)$$

where the quark Green's function is now given by

$$\mathcal{G}^{-1} = \begin{pmatrix} i\partial\!\!\!/ - m_i & -\Delta C \\ -C\tilde{\Delta} & (i\partial\!\!\!/ - m_i)^T \end{pmatrix}. \qquad (5.18)$$

Again, the quark loop is ultraviolet divergent and needs regularization. As in the previous chapters we will use Schwinger's proper time scheme after continuation of $\mathcal{G}$ to Euclidean space. For the purpose of this section we can neglect the imaginary part. The real part is then substituted by

$$\mathrm{ReTr} \log \mathcal{G}_E^{-1} \to -\frac{1}{2} \int_{1/\Lambda^2}^\infty \frac{ds}{s} \mathrm{Tr} \exp\left(-s(\mathcal{G}_E^{-1})^\dagger \mathcal{G}_E^{-1}\right). \qquad (5.19)$$

The Bethe–Salpeter equation for the diquarks can be derived along the lines described in Sect. 3.4. A little bit more care has to be used when taking into account the 2×2 matrix structure of the Nambu–Gorkov formalism. Expanding the action (5.17) up to second order in the diquark fields

$$\mathcal{A}_\Delta = \mathcal{A}_0 + \int \frac{d^4q}{(2\pi)^4} \, \Delta_\alpha^*(-q)(D_d^{-1})^{\alpha\beta}(q)\Delta_\beta(q) + \dots \qquad (5.20)$$

defines the diquark propagator D_d. The Bethe-Salpeter equation for diquarks is of a similar form to the one for mesons (3.62),

$$(D_d^{-1})^{\alpha\beta}(q^2 = -m_d^2) = \left(-\frac{3}{4g_2}\delta^{\alpha\beta} - \Pi_d^{\alpha\beta}(q^2 = -m_d^2)\right) = 0. \qquad (5.21)$$

For the case of a scalar diquark build of the two light flavors the polarization tensor is given in the isospin symmetric limit by

$$\Pi_d^{0^+}(q^2) = -q^2 \frac{1}{16\pi^2} \int_0^1 d\alpha \, \Gamma\left(0, \frac{m_u^2 + \alpha(1-\alpha)q^2}{\Lambda^2}\right) + \frac{\langle \bar{u}u \rangle}{6m_u}. \qquad (5.22)$$

The overall factor $1/2$ as compared to the meson case is due to different conventions for the (real) meson and (complex) diquark fields and has no influence on the pole position. Note also that the pion field was defined in sect. 3.4 as a dimensionless quantity whereas the diquark field has the usual dimension of a Bose field, namely the one of an energy. This also simply provides a common factor. On the other hand, the color trace is different for mesons and diquarks. Especially, if one compares pseudoscalar mesons to scalar diquarks in the Fierz symmetric case $g_2 = g_1$ one notes that the interaction in the diquark channel is a factor $1/3$ weaker than in the meson channel. This difference implies that the scalar diquarks are much more weakly bound than the pseudoscalar mesons. Only choosing $g_2 = 3g_1$ renders the scalar diquark mass equal to the pseudoscalar meson mass. Especially, in the chiral limit the light flavor scalar diquark is massless for $g_2 = 3g_1$. For smaller values of g_2 the scalar diquark mass is appreciably larger than the pion (kaon) mass. Nevertheless, it is still the lightest diquark. In fact, it turns out that only scalar and axialvector diquarks are bound for reasonable values of the coupling constant g_2.

Before giving values for the diquark masses we will discuss also electromagnetic properties of the light flavor diquarks. The electromagnetic form factors of the constituent diquarks are an important ingredient in the electromagnetic structure of baryons. To describe the electromagnetic interactions of diquarks, we couple an electromagnetic field to the quark fields by way of minimal substitution, $i\partial\!\!\!/ \to i\partial\!\!\!/ - QA_\mu\gamma^\mu$, where $Q = \frac{1}{2}e(\lambda^3 + \frac{1}{\sqrt{3}}\lambda^8)$ is the quark charge matrix. We then expand the effective action in the background of the electromagnetic field to second order in the diquark field and first order in the photon field:

$$
\mathcal{A}_\Delta^{(2,1)} = \int \frac{d^4q}{(2\pi)^4}\, \Delta_\alpha^*(-q)(D_d^{-1})^{\alpha\beta}(q)\Delta_\beta(q)
$$
$$
+ \int \frac{d^4p}{(2\pi)^4} \int \frac{d^4q}{(2\pi)^4} \Delta_\alpha^*(-p-\tfrac{1}{2}q)\Delta_\beta(p-\tfrac{1}{2}q)A_\mu(q)\mathcal{F}^{\alpha\beta\mu}(p,q),
$$

$$(5.23)$$

where $\mathcal{F}$ is the electromagnetic vertex function. Here, again the three–quark field is set to zero, $\chi = 0$, and the meson field is left at its vacuum value. On–shell diquark form factors are then obtained by evaluating the vertex function for appropriate incoming and outgoing four–momenta $p \pm \frac{1}{2}q$ on the diquark mass shell and normalizing the fields in eq. (5.23) to unit residue of the propagator. Note that the electromagnetic couplings of the diquark fields come entirely from the quark loop. In proper–time regularization corresponding expressions are again derived using standard techniques for the expansion of the time–ordered exponential, see Sect. 3.4. The electromagnetic properties of the light flavor scalar and axialvector diquarks, which are relevant to the description of baryons, have been calculated in ref. [WBAR93] and are displayed in Tab. 5.1 for different values of the effective diquark coupling constant, g_2. Here, the coupling in the meson channel, g_1, is determined by the constituent quark mass through the gap equation, the cutoff is fixed by fitting the pion decay constant, and the current mass is determined from the pion mass. Diquark couplings of $g_2/g_1 \sim 2$ are required to reproduce the masses of the spin-$\frac{1}{2}$ baryons if quark exchange is included [BAR92]. Similar values are also required to obtain sufficiently bound axial diquarks. For completeness the diquark masses are also shown in Tab. 5.1.

To obtain the electromagnetic form factor for an on–shell scalar diquark of mass m_{0^+}, one evaluates the vertex function $\mathcal{F}$ for incoming and outgoing momenta $p \pm \frac{1}{2}q$ such that $(p \pm \frac{1}{2}q)^2 = m_{0^+}^2$. Note that this leads to the condition $p \cdot q = 0$. On the mass shell, the vertex function is transverse to the photon momentum,

$$
\mathcal{F}_{0^+}^\mu(p,q) = 2QZ_{0^+}f_{0^+}(q^2)\,p^\mu \tag{5.24}
$$

Table 5.1. The masses of the scalar ud– and the axial vector uu–diquark, m_{0+} and m_{1+}, and their electromagnetic charge radii, for various values of the effective diquark coupling constant, g_2, and $m_u = 400\text{MeV}$. Also shown is the magnetic moment of the axial uu–diquark, μ_{1+}, in units of $2Q_u e/2m_u$. (Taken from ref. [WBAR93].)

g_2/g_1	m_{0+}/MeV	m_{1+}/MeV	$\langle r^2\rangle_{0+}^{1/2}/\text{fm}$	$\langle r^2\rangle_{1+}^{1/2}/\text{fm}$	μ_{1+}
1	700	–	0.55	–	–
1.5	573	–	0.49	–	–
2	449	787	0.48	1.08	0.95
2.5	318	764	0.48	0.75	0.91
3	140	741	0.48	0.65	0.88

where the normalization constant Z_{0+} is defined as $Z_{0+} = \frac{\partial \Pi_d^{0+}(q^2)}{\partial q^2}\big|_{q^2=m_{0+}^2}$.

The scalar diquark form factor as a function of the photon momentum, q^2, is almost identical to the one of a pion [WBAR93]. In particular, $f_{0+}(0) = 1$, $i.e.$ the total charge is conserved as a consequence of the gauge–invariant regularization. The scalar diquark r.m.s. charge radius, $\langle r^2\rangle_{0+} = -6\,\partial/\partial q^2 f_{0+}|_{q^2=0}$, is given in Tab. 5.1. As a matter of fact, the only difference between the diquark and the meson vertex function is the sign of the quark charges and a factor N_c multiplying the quark loop. However, since the same factor N_c occurs also in the meson propagator, this factor cancels when considering the normalized on–shell pion form factor. Thus, in the isospin limit the pion form factor is given by the expression for the diquark form factor evaluated with $m_{0+}^2 \rightarrow m_\pi^2$.

One may consider the scalar diquark (or pion) r.m.s. charge radius in Tab. 5.1 as a function of the bound state mass. For strongly bound diquarks the charge radius is essentially independent of the diquark mass and thus practically identical to that of the pion. It is therefore estimated very well by the result of a gradient expansion which reads

$$\langle r^2\rangle_{0+} = \frac{N_c}{4\pi^2 f_\pi^2} \exp(-m^2/\Lambda^2). \tag{5.25}$$

In the other limiting case $m_{0+} \rightarrow 2m_u$, where the diquark would become unbound, the charge radius grows like $\langle r^2\rangle_{0+} \sim (m_{0+} - 2m_u)^{-1}$. This is the behavior expected for a weakly bound state below a continuum threshold. Such a behaviour would be absent in a model which incorporates quark confinement.

We will close this section with a few remarks on the axial vector diquark $(\Gamma^\rho = i\gamma^\rho/\sqrt{2})$. With the gauge–invariant proper–time regularization and in the isospin symmetric case, the axial vector diquark polarization operator is transverse,

$$\Pi_{1+}^{\rho\sigma}(q^2) = \frac{q^2}{2g^2(q^2)}(\delta^{\rho\sigma} - \frac{q^\rho q^\sigma}{q^2}), \tag{5.26}$$

with

$$\frac{1}{g^2(q^2)} = \frac{1}{4\pi^2} \int_0^1 d\alpha\, \alpha(1-\alpha)\, \Gamma\left(0, \frac{m_u^2 + \alpha(1-\alpha)q^2}{\Lambda^2}\right). \qquad (5.27)$$

The longitudinal part of eq.(5.21) is thus given entirely by $(3/2g_2)g^{\rho\sigma}$. Axial vector diquark masses are given in Tab. 5.1 for various values of g_2. In contrast to the scalar diquark the axial vector diquark is only weakly bound. The electromagnetic vertex function for on–shell axial vector diquarks of incoming and outgoing four-momenta $p \pm \frac{1}{2}q$, with $(p \pm \frac{1}{2}q)^2 = m_{1+}^2$, is of the form [LY62]

$$\mathcal{F}_{1+}^{\rho\sigma\mu}(p,q) = 2QZ_{1+}\left(2g^{\rho\sigma}p^\mu f_{1+}^{\rm e}(q^2) - [g^{\rho\mu}(p+\tfrac{1}{2}q)^\sigma + g^{\sigma\mu}(p-\tfrac{1}{2}q)^\rho]f_{1+}^{\rm m}(q^2)\right) \qquad (5.28)$$

where $f_{1+}^{\rm e,m}(q^2)$ are the normalized electric and magnetic form factors of the axial vector diquark, respectively. The normalization constant Z_{1+} is obtained as in the case of the scalar diquark as the derivative of the polarization tensor taken on the mass shell. In particular, $f_{1+}^{\rm e}(0) = 1$, i.e. charge is conserved, and $f_{1+}^{\rm m}(0)$ is the magnetic moment of the ij–axial vector diquark in units of $(Q_i + Q_j)e/2m_{1+}$ $(i,j=u,d)$. The charge radius of the axial diquark is larger than that of the scalar diquark, see Tab. 5.1. This fact is mainly due to the weaker binding of the axial vector diquark. However, even strongly bound axial vector diquarks would be larger than scalar diquarks of the same mass. In gradient expansion at $m_{0+} = m_{1+} = 0$ one finds $\langle r^2 \rangle_{1+} = \frac{27}{20}\langle r^2 \rangle_{0+}$. The axial diquark magnetic moment is shown in Tab. 5.1 in units of the sum of the constituent quark magnetic moments, $(Q_i + Q_j)e/2m$. The diquark magnetic moment is seen to be lowered compared to the non–relativistic value as a consequence of the diquark binding.

5.3 The Relativistic Faddeev Equation for Baryons

In this section we will use the diquarks to describe baryons. From the generating functional $W[\eta, \chi]$ (5.6) we can in principle evaluate all meson and baryon properties. Here we will focus on the baryon mass. We follow again ref. [Re90]. The baryon propagator (5.15) can be expressed with the help of the effective action (5.7) as

$$G_{\alpha\beta}^B[\eta](x, X; y, Y) = \left\langle \left(\frac{\delta\mathcal{A}_{\rm eff}^2[\eta, \Delta, \chi]}{\delta\bar{\chi}_\alpha(x, X)\delta\chi_\beta(y, Y)} \right)_{\chi=0} \right\rangle$$
$$- \left\langle \left(\frac{\delta\mathcal{A}_{\rm eff}[\eta, \Delta, \chi]}{\delta\bar{\chi}_\alpha(x, X)} \right)_{\chi=0} \right\rangle \left\langle \left(\frac{\delta\mathcal{A}_{\rm eff}[\eta, \Delta, \chi]}{\delta\chi_\beta(y, Y)} \right)_{\chi=0} \right\rangle \qquad (5.29)$$

where

$$\langle \cdots \rangle = \frac{\int \mathcal{D}\Delta^* \mathcal{D}\Delta \cdots \exp\mathcal{A}_{\rm eff}[\eta, \Delta, \chi = 0]}{\int \mathcal{D}\Delta^* \mathcal{D}\Delta \exp\mathcal{A}_{\rm eff}[\eta, \Delta, \chi = 0]} \qquad (5.30)$$

denotes the functional averaging over the diquark fields. The unlinked terms

$$\left\langle\left(\frac{\delta\mathcal{A}_{\text{eff}}[\eta,\Delta,\chi]}{\delta\bar{\chi}_\alpha(x,X)}\right)_{\chi=0}\right\rangle \quad \text{and} \quad \left\langle\left(\frac{\delta\mathcal{A}_{\text{eff}}[\eta,\Delta,\chi]}{\delta\chi_\beta(y,Y)}\right)_{\chi=0}\right\rangle \qquad (5.31)$$

vanish since χ occurs only in the combination $\bar{\chi}\chi$, both in the baryon propagator G_B (5.15) and the effective action $\mathcal{A}_{\text{eff}}$ (5.7). To determine the baryon propagator one has to integrate out the diquark fields Δ and Δ^*. We will do this by the method of stationary phases. Note that the effective action $\mathcal{A}_{\text{eff}}$ has a stationary point at $\Delta = 0$, which is stable in the absence of quark matter.

Using the explicit form of the effective action $\mathcal{A}_{\text{eff}}$ (5.7) the baryon propagator can be written as

$$G_B[\eta](x,X;y,Y)_{\alpha\beta} = \left(\frac{3}{2g_2}\right)^2$$
$$\left\langle G_{11}[\eta,\Delta](x,y)\left(\Delta_\alpha(X)\Delta_\beta^*(Y) - i\frac{4}{3}g_2 g_{\alpha\beta}\delta(X-Y)\right)\right\rangle$$

$$(5.32)$$

where

$$G_{11}[\eta,\Delta] = \left(1 - G[\eta]\Delta\tilde{G}[\eta]\tilde{\Delta}\right)^{-1} G[\eta], \quad \tilde{G}[\eta] = C^\dagger G[\eta]C \qquad (5.33)$$

is the normal part of the quark Green's function interacting with the diquark field Δ. Furthermore, $g_{\alpha\beta}$ denotes the Kronecker symbol $\delta_{\alpha\beta}$ for all indices except the Dirac indices where it stands for the metric tensor. In the evaluation of the averages $\langle\cdots\rangle$ we will use the Gaussian approximation. We therefore resort to a perturbative expansion of G_{11} in powers of Δ

$$G_{11}[\eta,\Delta] = \sum_{n=0}^\infty \left(G[\eta]\Delta\tilde{G}[\eta]\tilde{\Delta}\right)^n G[\eta]. \qquad (5.34)$$

Using

$$\langle\Delta_\alpha(X)\Delta_\beta^*(Y)\rangle = (D_d)_{\alpha\beta}(X,Y) \qquad (5.35)$$

in lowest order the baryon propagator is given by

$$G_B[\eta](x,X;y,Y)_{\alpha\beta} = \left(\frac{3}{2g_2}\right)^2 G[\eta](x,y)\,i(\tilde{D}_d)_{\alpha\beta}[\eta](X,Y) \qquad (5.36)$$

where

$$i(\tilde{D}_d)_{\alpha\beta}[\eta](X,Y) = i(D_d)_{\alpha\beta}[\eta](X,Y) - i\frac{4}{3}g_2 g_{\alpha\beta}\delta(X-Y). \qquad (5.37)$$

A straightforward explicit evaluation of the propagator D_d (5.21) shows that it contains a (non–propagating) contact term which is precisely canceled by the second term in (5.37). The zero–order baryon propagator (5.36) describes the

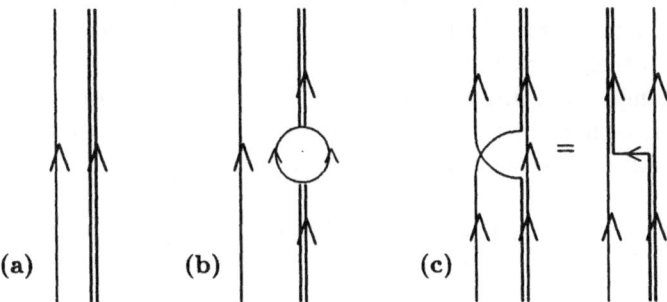

Fig. 5.1. (a) The unperturbated baryon propagator describing the independent propagation of a quark and a diquark. By construction of the diquark propagator this diagram includes already the diagram shown in (b). (c) Exchange diagram to (b).

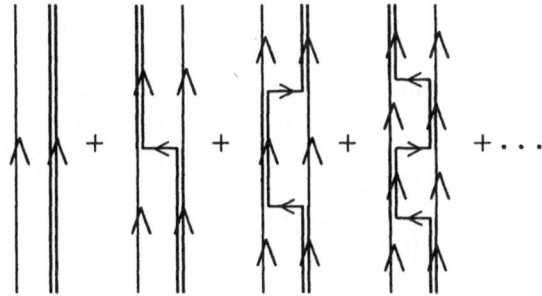

Fig. 5.2. Exchange diagrams to the leading order diagram shown in Fig. 5.1a.

independent propagation of a diquark and a quark, both in the background of the fluctuating meson field η, see Fig. 5.1.

It is, however, not consistent to stick to this zero order approximation for the following reason: The diquark field (propagator) has been constructed by summing partially all diquark bubbles. Thus the leading order baryon diagram Fig. 5.1a contains in fact already all the diagrams arising from the diquark correlations. The leading order diagram of this class is shown in Fig. 5.1b. This diagram contains three intermediate quarks which should be in a completely antisymmetric state. But since the diquark was constructed irrespectively of the presence of the third quark this intermediate three–quark state will in general not fulfill the Pauli principle. To obtain a consistent description of the baryons which takes fully care of the Pauli principle, in addition to the leading order diagram, Fig. 5.1a, one has to include at least the corresponding exchange diagrams shown in Fig. 5.2. Due to the color of the quarks these exchange correlations are attractive and, as we will see later, bind diquark and quark into baryons. Readers familiar with Pauli repulsion effects in nuclear many–body systems may be confused at that point, but antisymmetry in color requires symmetrization of all other quantum numbers. The leading order exchange diagram corresponding to the leading order diquark bubble 5.1b is shown in 5.1c.

The exchange diagrams shown in Fig. 5.2 are all generated when the expansion (5.34) is inserted into the expression (5.32) for the baryon propagator. Keeping only these diagrams it is then given by

$$G_B = \left(\frac{3}{2g_2}\right)^2 \left((G_B^{0-1} - H)^{-1} + \delta G_B^0\right) \tag{5.38}$$

where

$$G_B^0[\eta] = G[\eta] i D_d[\eta] \tag{5.39}$$

is the independent quark–diquark propagator shown in Fig. 5.1a. The quantity

$$H_{\alpha\beta}(x, X; y, Y) = \Sigma_\beta \tilde{G}(x, y) \Sigma_\alpha \delta(X - y) \delta(Y - x) \tag{5.40}$$

is the quark exchange vertex shown in Fig. 5.1c. $(\delta G_B^0)_{\alpha\beta}$ removes the contact term contained in the diquark propagator from the leading order term $G^0 D_d$.

The baryon states are defined by the poles of the baryon Green function $G_B[\eta]$. Since the contact term δG_B cannot develop a non–trivial pole the baryon states are given by the solution of the Faddeev type equation $G_B^{-1}\psi = 0$. We are interested in the poles of this propagator in the background of the vacuum configuration of the meson fields, which defines the baryon masses. This Faddeev type equation is for its numerical treatment best expressed as

$$\psi(p, q_1) + \int \frac{d^4 q_2}{(2\pi)^4} L(p, q_1, q_2) \psi(p, q_2) = \lambda(p) \psi(p, q_1). \tag{5.41}$$

Here $\psi(p, q)$ is the baryon field and the eigenvalue $\lambda(p)$ vanishes on the baryon mass shell $p^2 = -m_B^2$. The integral kernel L can be written as a product

$$L^{\alpha\gamma}(p, q_1, q_2) = (G_B^0)^{\alpha\beta}(p, q_1) H^{\beta\gamma}(q_1, q_2) \tag{5.42}$$

where

$$(G_B^0)^{\alpha\beta}(p, q_1) = i G(p/2 + q_1) D_d^{\alpha\beta}(p/2 - q_1) \tag{5.43}$$

describes the unperturbed propagation of a quark and a diquark (see the first diagram of Fig. 5.1 and eq. (5.39)) and

$$H^{\beta\gamma}(q_1, q_2) = \Sigma^\gamma G(-q_1 - q_2) \Sigma^\beta \tag{5.44}$$

is the quark exchange vertex containing Dirac, flavor and colour matrices, see eq. (5.40), as well as the propagator of the exchanged quark.

The color structure of the baryon self–energy H (5.44) is given by

$$H^{\alpha\beta} \sim \frac{i\epsilon_{CD}^A}{\sqrt{2}} \frac{i\epsilon_{DE}^B}{\sqrt{2}} = -P_{(1)}{}_{CE}^{AB} + \frac{1}{2} P_{(8)}{}_{CE}^{AB} \tag{5.45}$$

where $P_{(1)CE}^{AB} = \delta_C^A \delta_E^B / 3$ and $P_{(8)CE}^{AB} = \delta^{AB} \delta_{CE} - P_{(1)CE}^{AB}$ denote the color singlet and octet projectors, respectively. Thus, independent from the flavor or Dirac content, the self–energy H induces no mixing between color singlet and octet states, and the sign of the self–energy is different in both cases. Thereby we obtain the encouraging result that, if the diquark–quark interaction is attractive in the color singlet baryon channel (as we shall see below), the interaction is repulsive for the unphysical color octet states. As the quark and the diquark propagator are diagonal in color the Faddeev equation (5.41) is easily decomposed into color singlet and octet parts. We are only interested in the physical color singlet state (anyhow, as the baryon self–energy is negative in the octet states these are not bound). Since the color indices are completely fixed in the singlet channel we will simply omit them in the following.

For pedagogical reasons we will restrict ourselves to scalar diquarks as building blocks of baryons. As scalar diquarks are much lighter than the axialvector diquarks they are probably the most important diquark correlations in baryons. As discussed in the previous section the Pauli principle for diquarks necessitates the scalar diquarks to be in a flavor antitriplet state. Hence scalar diquarks are build up by two constituent quarks of different flavor. Therefore, baryons containing a scalar diquark are flavor singlet or octet states consisting at least of two different quark flavors. Note further that neglecting other then scalar diquarks the spin of the baryons is entirely carried by the third valence quark. The Dirac spinor ψ is then given by the spinor of the valence quark times a scalar function $(\bar{q}^C i \gamma^5 q)$. Thus the restriction to scalar diquark correlations yields only baryons with spin $1/2$.

The Faddeev equation (5.41) is a non–separable integral equation. Instead of solving it by brute force numerically it is probably more instructive to use transparent approximations. An approximation simplifying the Faddeev equation enormously is to discard the momentum dependence of the exchanged quark

$$G_i(-q_1 - q_2) = \frac{1}{-\not{q}_1 - \not{q}_2 - m_i} \rightarrow \frac{1}{m_i}. \tag{5.46}$$

We will refer to this approximation as static approximation. Note that it becomes accurate if the exchanged quark is infinitely heavy. In this approximation the Faddeev equation (5.41) reduces to a Dirac equation with momentum dependent coefficients

$$\left(\not{p} A(p^2) + B(p^2) - 1 \right) \phi(p)|_{p^2 = -m_B^2} = 0 \tag{5.47}$$

where

$$\phi(p) = \int \frac{d^4 q}{(2\pi)^4} \psi(p, q). \tag{5.48}$$

For the case of two flavors the quantities A and B are given by

$$A = \frac{1}{2} I^0 + I^1, \quad B = m_u I^0,$$

$$I^s = \frac{1}{m_u} \int \frac{d^4q}{(2\pi)^4} \left(\frac{pq}{p^2}\right)^s \frac{D_d(p/2-q)}{(p/2+q)^2 - m_u^2} .$$
(5.49)

The integrals appearing in the Dirac equation (5.47) have to be cutoff by some procedure. Note that this need for regularization stems from the static approximation (5.46), *i.e.* the integral in the Faddeev equation (5.41) is in principle UV finite.

Table 5.2. Masses (in MeV) for spin 1/2 baryons in the static approximation in comparison to the experimental values (last row). g_2 is chosen such that $m_p = 938$MeV. (Taken from ref. [BAR92]).

		$m_u = 350$MeV	$m_u = 400$MeV	$m_u = 450$MeV	exp.
p	(uud)	938	938	938	938
n	(udd)	941	938	938	939
Λ^0	(uds)	1097	1068	1061	1116
Σ^0	(uds)	—	1209	1233	1193
Σ^+	(uus)	—	1205	1229	1189
Σ^-	(dds)	—	1214	1236	1197
Ξ^0	(uss)	1319	1292	1285	1315
Ξ^-	(dss)	1328	1300	1292	1321

For three flavors the masses of the spin 1/2 baryons in the static approximation (5.46) to the exchanged quark have been calculated in ref. [BAR92] and are displayed in Tab. 5.2 for the three cases m_u=350, 400 and 450 MeV. For $m_u = 350$MeV the Σ baryon is still unbound and becomes bound only for larger constituent masses, despite that g_2 has been increased to keep the proton mass fixed. The us or the ds diquark is bound. It is the quark exchange in the static approximation which is to weak. Besides this problem the calculated baryon masses are in good agreement with the experimental data for up constituent masses between 350 and 400 MeV. For the flavor diagonal baryons, *i.e.* the one with uds quark content, flavor mixing occurs in the Faddeev equation leading to a coupled system of equations for mass eigenstates. In a range of constituent masses 350 MeV $\leq m_u <$ 400 MeV we obtain only one bound state out of three possible flavor diagonal states whereas for $m_u \geq$ 400 MeV we found two bound baryons out of three possible states. The latter situation agrees with experiment as Λ^0 and Σ^0 are the only measured uds–baryons. Furthermore, there is one state almost degenerate to $\Sigma^\pm$ which we then naturally identify with Σ^0 whereas the other bound state has always lower mass and is therefore interpreted as Λ^0.

Summary

Using functional integral techniques the quark formulation of QFD (2.24) has been transformed into a theory of hadrons, *i.e.* mesons and baryons, see eq. (5.5). However, the auxiliary three–quark field which had to be introduced at an intermediate step can be integrated out only approximately yielding (5.16). The resulting theory contains a term canceling anomalies which may, in principle, arise from the composite baryon field. This is a gratifying result because double counting of anomalies is avoided: Anomalies arise in this theory exclusively due to the fundamental fermions, the quarks.

In the process of functional integral hadronization one realizes the importance of diquarks. First, due to algebraic reasons: In a physical baryon every pair of quarks has to be in an antitriplet color state. Second, dynamically: Scalar and axialvector diquarks are bound objects within QFD. Furthermore, the Pauli principle requires the scalar diquark to be antisymmetric in flavor, the axialvector diquark to be symmetric.

In a calculation where only the scalar diquark and the kinematical diquark–quark exchange correlations have been taken into account the latter binds quarks and diquarks into baryons. In agreement with experiment only eight bound baryons out of nine possible states have been found. In the used static approximation where the exchanged quark is treated as infinitely heavy the Faddeev equation is reduced to the Dirac type equation (5.47) with momentum dependent mass term.

Obviously, the investigations reported in this chapter are the basis for further work. Inclusion of axialvector diquarks, and correspondingly the description of spin $\frac{3}{2}$ baryons, calculations of *e.g.* electroweak decays and form factors are highly wanted. These issues are currently under investigation.

Chapter 6

On the Description of Baryons: An Outlook

In these Lecture Notes we have demonstrated that a model motivated from QCD, namely Quantum Flavor Dynamics (2.24), can describe hadrons extremely well. Here we have put more emphasis on the description of baryons than on the one of mesons. Surprisingly, there are two possibilities to build baryons. One, which is the only one in the limit of a large number of colors, is based on the picture of baryons as chiral solitons. A special feature of this part of Quantum Flavor Dynamics is that Witten's conjecture [Wi79] is fulfilled if all possible Lorentz tensors of the meson fields are included. On the other hand, for a finite number of colors the baryon may be introduced as a ordinary bound state of three valence quarks. So far, investigations have been restricted to the case that meson fields are frozen to its vacuum expectation value. A complete investigation of Quantum Flavor Dynamics would require the explicit baryon field as introduced in Sect. 5.1 as well as the non–perturbative (solitonic) meson fields, especially the hedgehog (4.3) for the pseudoscalar field.

Allowing in the expression (5.5), or in its approximation (5.16), non–trivial meson and baryon field configurations is certainly an exceedingly difficult task. On the other hand, it is a clearly defined formal procedure. Thus approximations may be improved step by step taking into account more and more the hybrid nature of baryons.

In a first step towards such a complete description of baryons within Quantum Flavor Dynamics, for the background meson field in the diquark action (5.17) a solitonic field of the hedgehog type (4.3) (instead of the mesonic vacuum expectation values) has been used [ZARW]. This amounts to use the quark Green' s function (for two flavors in the isospin limit)

$$\mathcal{G}^{-1} = \begin{pmatrix} i\partial\!\!\!/ - m_u U^{\gamma_5} & -\Delta C \\ -C\tilde{\Delta} & (i\partial\!\!\!/ - m_u U^{\gamma_5})^T \end{pmatrix}. \qquad (6.1)$$

As the hedgehog *ansatz*

$$U(\boldsymbol{x}) = \exp\!\big(i\boldsymbol{\tau} \cdot \hat{\mathbf{r}}\,\Theta(r)\big) \tag{6.2}$$

is spherically symmetric it is convenient to expand the diquark propagator (5.21) in a grand spin basis similar to the treatment of mesonic fluctuations off the soliton, see Sect. 4.6. The resulting Bethe–Salpeter equation for diquarks resembles formally the one for the pions, eq. (4.89). However, the corresponding kernels are quite complicated. Due to the symmetries of the hedgehog it proves advantageous not to use the charge conjugated quark Green's function $\tilde{G}$ (5.33) but the G–parity transformed one. This then allows a decomposition of the kernels into grand spin and parity channels. Furthermore, as in the meson case the kernels can be expressed as mode sums involving eigenfunctions and eigenvalues of the Dirac Hamiltonian (4.10).

In the preliminary investigations of [ZARW] the *ansatz* (4.9) for the chiral angle has been used. The scalar diquark mass, or equivalently the scalar diquark binding energy, has been studied as a function of the soliton size a. Obviously, in the limit $a \to 0$ (no soliton) the continuum result is rediscovered. As a matter of fact, the (small) deviations from the value calculated in a plane wave basis give some estimate of finite size effects when using the grand spin basis in a finite spherical box. Considering the diquark mass as a function of a one has to distinguish between two effects: First, the valence quark energy decreases continueously for increasing a (see fig. 4.1), *i.e.* for a bound diquark the diquark mass decreases drastically already due to this effect. Second, the genuine diquark binding energy which for the lowest mass diquark might be defined as $2\epsilon_{\text{val}} - m_{0^+}$ changes with a. In the investigations of [ZARW] it was found that the binding energy increases for small values of a up to approximately twice its vacuum value whereas it decreases for larger a. Most surprisingly, the maximum of the diquark binding energy and therefore the minimum of the diquark mass occurs for the same value of a (within numerical errors) where the soliton energy has its minimum. At this point it is certainly premature to draw hard conclusions. Taken seriously, this preliminary result, however, indicates the following scenario: The soliton is not changed in shape by diquark correlations but the total energy of the baryon is decreased by a few hundred MeV due to genuine binding effects between two quarks. This result should provide enough motivation for investigations of the Faddeev equation in the soliton background which may shed some light on the fact that both, soliton and quark models of baryons, are successful.

Appendix A
Covariant Gradient Expansion

The method we are going to discuss here is a relativistic covariant generalization of the Wigner–Kirkwood expansion (see Sect. 13.2.2 of ref. [RS80] and references therein). We follow here ref. [ER86]. Our aim is a long wavelength expansion for a momentum–dependent operator $A(x^\mu, \partial_\mu)$. Note that $[x^\mu, \partial_\nu] = \delta^\mu_\nu \neq 0$ so we have to be careful expanding the operator A. In a first step we define the "heat kernel"

$$K(\tau) = e^{-\tau A} \tag{A.1}$$

or in matrix notation

$$\langle x|K(\tau)|y\rangle = \langle x|e^{-\tau A}|y\rangle. \tag{A.2}$$

This operator fulfills the heat or diffusion equation for imaginary times τ

$$\frac{\partial}{\partial \tau}K(\tau) + AK(\tau) = 0 \tag{A.3}$$

with the boundary condition $K(\tau = 0) = 1$. In order to apply perturbation theory we split the operator A into two parts

$$A = A_0 + V \tag{A.4}$$

and define the zeroth order heat kernel as

$$K_0(\tau) = e^{-\tau A_0}. \tag{A.5}$$

We use the ansatz

$$K = K_0 \star H \tag{A.6}$$

where the product " $\star$ " denotes multiplication of matrix elements

$$\langle x|K(\tau)|y\rangle = \langle x|K_0(\tau)|y\rangle\langle x|H(\tau)|y\rangle. \tag{A.7}$$

Eqs. (A.1), (A.5) and (A.7) imply the boundary condition

$$\langle x|H(0)|y\rangle = 1. \tag{A.8}$$

The operator we want to expand is $\not{D}^\dagger \not{D}$ (3.83). As zeroth order term we choose the inverse propagator of a free Klein–Gordon field

$$A_0 = \partial_\mu \partial^\mu + \mu^2. \tag{A.9}$$

Then one obtains

$$\langle x|K_0(\tau)|y\rangle = \frac{1}{(4\pi\tau)^{D/2}} \exp\left(\frac{(x-y)^2}{4\tau} - \mu^2\tau\right) \tag{A.10}$$

where D is the number of space–time dimensions. Note that $(x-y)^2$ is negative semi–definite because we use negative Euclidean matrix $g_{\mu\nu} = -\delta_{\mu\nu}$. Furthermore,

$$\lim_{\tau\to 0} \langle x|K_0(\tau)|y\rangle = \delta^{(D)}(x-y). \tag{A.11}$$

Using the ansatz (A.6) in the heat equation (A.3) and exploiting the specific form of A (3.83) and A_0 (A.9) one obtains the following equation for the matrix elements of $H(\tau)$

$$\left(\frac{\partial}{\partial\tau} + \frac{1}{\tau}(x-y)_\mu d^\mu + d_\mu d^\mu + a\right) \langle x|H(\tau)|y\rangle = 0 \tag{A.12}$$

where $d_\mu = \partial/\partial x^\mu + \Gamma_\mu$. Expanding this matrix element in powers of the proper time τ

$$\langle x|H(\tau)|y\rangle = \sum_{k=0}^{\infty} h_K(x,y)\tau^K \tag{A.13}$$

and using this in eq. (A.12) one derives the recursion relation

$$(n+1+(x-y)_\mu d^\mu)\, h_{n+1}(x,y) + (a + d_\mu d^\mu)\, h_n(x,y) = 0 \tag{A.14}$$

with the starting condition

$$(x-y)_\mu d^\mu\, h_0(x,y) = 0. \tag{A.15}$$

Inserting (A.7) and (A.13) in the proper time regularization we obtain

$$\mathcal{A}_F^R = \frac{1}{2}\mathrm{Tr}\log \not{D}^\dagger \not{D}$$
$$= \frac{1}{2}\sum_{k=0}^{\infty}\int \frac{d\tau}{\tau}\, \tau^k\, \mathrm{Tr}(K_0(\tau)h_n) \tag{A.16}$$

where

$$\mathrm{Tr}(K_0(\tau)h_n) = \mathrm{tr}\int d^4x\, \langle x|K_0(\tau)|x\rangle h_n(x,x) \tag{A.17}$$

and "tr" denotes the trace over internal indices (Dirac, color and flavor). As for $D = 4$ (see eq. (A.10))

$$\langle x|K_0(\tau)|x\rangle = \frac{1}{(4\pi\tau)^2}e^{-\tau\mu^2} \tag{A.18}$$

we obtain

$$A_F^R = -\frac{1}{2}\sum_{k=0}^{\infty}\Gamma(\frac{\mu^2}{\Lambda^2}, k-2)\mathrm{Tr}(h_k). \tag{A.19}$$

Thus, in order to evaluate A_F^R we have to calculate the operators h_n explicitly. For $x \to 0$

$$\Gamma(-k, x) \sim \frac{1}{x^k}, \qquad \Gamma(0, x) \sim -\log x \tag{A.20}$$

whereas $\Gamma(n > 0, x = 0)$ is finite. For small μ^2/Λ^2 we therefore need only the first three terms in the expansion (A.19). Using $x = y$ the recursion formula (A.14) reads

$$(n+1)h_{n+1}(x, x) + d_\mu d^\mu h_n(x, y)|_{x=y} + ah_n(x, x) = 0 \tag{A.21}$$

From the boundary condition at $\tau = 0$ we know that

$$h_0(x, x) = 1. \tag{A.22}$$

For $n = 0$ eq. (A.21) reads

$$h_1(x, x) = -d_\mu d^\mu h_0(x, y)|_{x=y} - ah_0(x, x) \tag{A.23}$$

Acting with the operator d_ν on the boundary condition (A.15) we get

$$0 = d_\nu[(x - y)_\mu d^\mu h_0(x, y)] = d_\nu h_0(x, y) + (x - y)_\mu d_\nu d^\mu h_0(x, y) \tag{A.24}$$

and therefore

$$d_\nu h_0(x, y)|_{x=y} = 0. \tag{A.25}$$

Acting a second time with d_ν on the boundary condition

$$d^\nu d_\nu((x - y)_\mu d^\mu h_0(x, y)) =$$
$$d^\nu d_\nu h_0(x, y) + d_\nu d^\nu h_0(x, y) + (x - y)_\mu d^\nu d_\nu d^\mu h_0(x, y) = 0$$
$$\tag{A.26}$$

we conclude that

$$d^\nu d_\nu h_0(x, y)|_{x=y} = 0. \tag{A.27}$$

Using (A.22) we obtain the coefficient $h_1(x, x)$,

$$h_1(x, x) = -a. \tag{A.28}$$

Without proof we state the result for $h_2(x, x)$

$$h_2(x, x) = \frac{1}{2}a^2 + \frac{1}{6}[d_\mu, [d^\mu, a]] + \frac{1}{12}\Gamma_{\mu\nu}\Gamma^{\mu\nu}. \tag{A.29}$$

Appendix B
Topological Properties of the Chiral Field

We are considering time–independent, *i.e.* static, chiral fields

$$U(x) = e^{i\Theta(x)}, \quad \Theta(x) = \tau\Theta(x) \tag{B.1}$$

which are elements of the SU(2) flavor (isospin) group. The group manifold of SU(2) is the unit sphere in four dimensions, S^3. This becomes obvious if one parametrizes an element of the group $A \in$ SU(2) as

$$A = a_0 + a\tau. \tag{B.2}$$

The conditions $\det A = 1$ and $A^\dagger A = 1$ which are equivalent to $A \in$ SU(2) then require

$$a_0^2 + a^2 = 1 \tag{B.3}$$

or phrased in words: With the constraint (B.3) (a_0, a) define the unit sphere S^3. Therefore $U(x)$ describes a map from $\mathbb{R}^3 (\ni \mathbf{x})$ into S^3. If we are restricting ourselves to field configurations of finite energy we have to require

$$U(x) \to const. \quad \text{for} \quad |\mathbf{x}| \to \infty \tag{B.4}$$

because the leading term in gradient expansion would otherwise give an infinite contribution. Therefore all points at spatial infinity are equivalent for finite energy field configurations. Thus, in the treatment of the map $U(x)$ we may identify all points at spatial infinity. This compactifies the manifold $\mathbb{R}^3$ to S^3,

$$U : \mathbf{x} \in S^3_{spatial} \to S^3_{isospatial} \cdot \tag{B.5}$$

A simpler example of compactification is given by identifying the two end points of the line of real numbers: $\mathbb{R} \to S^1$. We will use this example because it has the benefit that the corresponding maps may be drawn on a sheet of paper.

Such maps are classsified by the so–called winding number which simply counts how often a map "wraps" the base manifold onto the target manifold.

106

Stated more precisely, the set of maps U is divided into equivalence classes called homotopy classes. Two maps U_1 and U_2 belonging to the same homotopy class may be continueously deformed until they are identical. In order to phrase it formally correct: There exists at least one relation called homotopy $U(x,s)$, $s \in [0,1]$ such that $U(x,0) = U_1(x)$, $U(x,1) = U_2(x)$ and $U(x,s)$ is continueous with respect to the parameter s. For topologically non–trivial maps there exist several homotopy classes. In the case $S^n \to S^m$ they form groups called $\Pi_n(S^m)$. In the two examples under consideration this group is isomorphic to the group of the integer numbers

$$\Pi_n(S^n) = \Pi_1(S^1) = \Pi_3(S^3) = \mathcal{Z} = \{0, \pm 1, \pm 2, \ldots\}. \tag{B.6}$$

The integer characterizing the homotopy class is the winding number. For the case of $S^1 \to S^1$ a few maps with different winding numbers are shown in Figure B.1. Obviously, this figures make the name "winding number" self–evident.

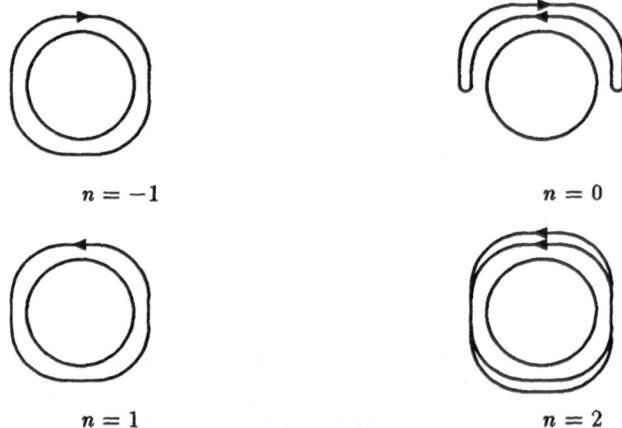

Fig. B.1. Maps $S^1 \to S^1$ with their corresponding winding numbers n.

We are going to proof now that winding number is given by the expression (3.149). Using the definition of the topological current (3.142) we obtain

$$N_B = \int d^3x \ B^0(x) = \frac{1}{24\pi^2} \ \epsilon^{ijk} \int d^3x \ \text{tr}(L_i L_j L_k) \ . \tag{B.7}$$

As $U(x) = U(\Theta(x))$ we use the chain rule to rewrite L_i as

$$L_i = U^\dagger \partial_i U = \partial_i \Theta^a L_a \ , \quad L_a = U^\dagger \frac{\partial U}{\partial \Theta^a}. \tag{B.8}$$

The baryon number can then be expressed as

$$N_B = \frac{1}{24\pi^2}\epsilon^{ijk}\int d^3x\ \partial_i\Theta^a\partial_j\Theta^b\partial_k\Theta^c\,\text{tr}L_aL_bL_c$$

$$= \frac{1}{24\pi^2}\int d^3x\ \epsilon^{abc}\det\left(\frac{\partial\Theta^d}{\partial x^l}\right)\text{tr}L_aL_bL_c$$

$$= \frac{1}{24\pi^2}\int d^3x\ \det\left(\frac{\partial\Theta^d}{\partial x^l}\right)\frac{1}{2}\epsilon^{abc}\text{tr}[L_a, L_b]L_c\ . \tag{B.9}$$

The determinant

$$\det\left(\frac{\partial\Theta^d}{\partial x^l}\right) \tag{B.10}$$

is the Jacobian for the transformation of integration variables from x to Θ. For topologically non–trivial maps with winding number n the integral measure is given by

$$d^3x\ \det\left(\frac{\partial\Theta^d}{\partial x^l}\right) = n\ d^3\Theta = n\ d\Theta^1 d\Theta^2 d\Theta^3. \tag{B.11}$$

Therefore we obtain

$$N_B = \frac{n}{48\pi^2}\int_{S^3} d^3\Theta\ \epsilon^{abc}\text{tr}[L_a, L_b]L_c$$

$$= \frac{n}{8\pi^2}\int_{S^3} d^3\Theta\ \text{tr}[L_1, L_2]L_3. \tag{B.12}$$

In order to paramatrize the unit sphere S^3 we choose Euler angles

$$U(x) = e^{i\Theta^1\tau_3}e^{i\Theta^2\tau_2}e^{i\Theta^3\tau_3}. \tag{B.13}$$

This leads to

$$\text{tr}[L_1, L_2]L_3 = \text{tr}[L_3, L_1]L_2 = \frac{1}{2}\sin\Theta^2 \tag{B.14}$$

Hereby Θ^2 denotes the second isospin component of Θ. Now it is easy to evaluate the integral (The factor 2 in front of the integral is part of the measure. This extra factor accounts for the fact that the Euler parametrization describes only proper rotations.)

$$N_B = \frac{n}{8\pi^2}2\int_0^{2\pi} d\Theta^1\int_0^{2\pi} d\Theta^3\int_0^\pi d\Theta^2\frac{1}{2}\sin\Theta^2$$

$$= n. \qquad \text{q.e.d.} \tag{B.15}$$

Appendix C
Equations of Motion for the NJL Soliton Meson Profiles

As the eigenfunctions (4.35) occur only in certain combinations in the equations of motion it is convenient to define quark density matrices. The scalar quark density matrix $\rho(\boldsymbol{x}, \boldsymbol{y})$ may be decomposed into contributions due to valence and sea quarks:

$$\rho(\boldsymbol{x}, \boldsymbol{y}) = \rho_R^{\text{val}} + \rho_I^{\text{val}} + \rho_R^{\text{vac}} + \rho_I^{\text{vac}},$$

$$\rho_R^{\text{val}}(\boldsymbol{x}, \boldsymbol{y}) = \sum_\nu \left\{ \psi_\nu^R(\boldsymbol{x}) \bar{\psi}_\nu^R(\boldsymbol{y}) - \psi_\nu^I(\boldsymbol{x}) \bar{\psi}_\nu^I(\boldsymbol{y}) \right\} \eta_\nu,$$

$$\rho_I^{\text{val}}(\boldsymbol{x}, \boldsymbol{y}) = \sum_\nu \left\{ \psi_\nu^R(\boldsymbol{x}) \bar{\psi}_\nu^I(\boldsymbol{y}) + \psi_\nu^I(\boldsymbol{x}) \bar{\psi}_\nu^R(\boldsymbol{y}) \right\} \eta_\nu,$$

$$\rho_R^{\text{vac}}(\boldsymbol{x}, \boldsymbol{y}) = \sum_\nu \left\{ \psi_\nu^R(\boldsymbol{x}) \bar{\psi}_\nu^R(\boldsymbol{y}) - \psi_\nu^I(\boldsymbol{x}) \bar{\psi}_\nu^I(\boldsymbol{y}) \right\} f_R(\epsilon_\nu/\Lambda),$$

$$\rho_I^{\text{vac}}(\boldsymbol{x}, \boldsymbol{y}) = \sum_\nu \left\{ \psi_\nu^R(\boldsymbol{x}) \bar{\psi}_\nu^I(\boldsymbol{y}) + \psi_\nu^I(\boldsymbol{x}) \bar{\psi}_\nu^R(\boldsymbol{y}) \right\} f_I(\epsilon_\nu/\Lambda). \tag{C.1}$$

Similarly, the quark number (or baryon number) density matrix $b(\boldsymbol{x}, \boldsymbol{y})$ is written as

$$b(\boldsymbol{x}, \boldsymbol{y}) = b_R^{\text{val}} + b_I^{\text{val}} + b_R^{\text{vac}} + b_I^{\text{vac}},$$

$$b_R^{\text{val}}(\boldsymbol{x}, \boldsymbol{y}) = \sum_\nu \left\{ \psi_\nu^R(\boldsymbol{x}) \psi_\nu^{R\dagger}(\boldsymbol{y}) - \psi_\nu^I(\boldsymbol{x}) \psi_\nu^{I\dagger}(\boldsymbol{y}) \right\} \eta_\nu,$$

$$b_I^{\text{val}}(\boldsymbol{x}, \boldsymbol{y}) = -\sum_\nu \left\{ \psi_\nu^I(\boldsymbol{x}) \psi_\nu^{R\dagger}(\boldsymbol{y}) + \psi_\nu^R(\boldsymbol{x}) \psi_\nu^{I\dagger}(\boldsymbol{y}) \right\} \eta_\nu,$$

$$b_R^{\text{vac}}(\boldsymbol{x},\boldsymbol{y}) = \sum_\nu \Big\{ \psi_\nu^R(\boldsymbol{x})\psi_\nu^{R\dagger}(\boldsymbol{y}) - \psi_\nu^I(\boldsymbol{x})\psi_\nu^{I\dagger}(\boldsymbol{y}) \Big\} f_I(\epsilon_\nu/\Lambda),$$

$$b_I^{\text{vac}}(\boldsymbol{x},\boldsymbol{y}) = -\sum_\nu \Big\{ \psi_\nu^I(\boldsymbol{x})\psi_\nu^{R\dagger}(\boldsymbol{y}) + \psi_\nu^R(\boldsymbol{x})\psi_\nu^{I\dagger}(\boldsymbol{y}) \Big\} f_R(\epsilon_\nu/\Lambda). \qquad \text{(C.2)}$$

Note that the baryon density matrix $b(\boldsymbol{x},\boldsymbol{y})$ differs from the scalar density matrix $\rho(\boldsymbol{x},\boldsymbol{y})$ not only by the additional factor γ_0 but also by an exchange of the regulator functions f_R and f_I which are given as the derivatives of the energy functional with respect to the energy eigenvalue ϵ_ν:

$$f_R(\epsilon_\nu/\Lambda) = \begin{cases} -\frac{1}{2}\text{sign}(\epsilon_\nu^R)\mathcal{N}_\nu, & \mathcal{A}_I \quad \text{not reg.,} \\[2mm] -\frac{1}{2}\text{sign}(\epsilon_\nu^R)\mathcal{N}_\nu + \frac{1}{\sqrt{\pi}}(\epsilon_\nu^I/\Lambda)\exp\big(-(\epsilon_\nu^R/\Lambda)^2\big), & \mathcal{A}_I \quad \text{regularized} \end{cases} \qquad \text{(C.3)}$$

$$f_I(\epsilon_\nu/\Lambda) = -\frac{1}{2}\text{sign}(\epsilon_\nu^R)\begin{cases} 1, & \mathcal{A}_I \quad \text{not regularized} \\[2mm] \mathcal{N}_\nu & \mathcal{A}_I \quad \text{regularized} \end{cases} \qquad \text{(C.4)}$$

with the vacuum occupation number $\mathcal{N}_\nu$ being defined in (4.24). With these definitions the equations of motion (4.34) for the meson profiles $\phi, \Theta, \omega, G, F$ and H read

$$\phi(r) = \frac{m_0}{m}\cos\Theta(r) - \frac{m_0 N_c}{m_\pi^2 f_\pi^2}\text{tr}\int\frac{d\Omega}{4\pi}\Big(\cos\Theta(r) + i\gamma_5\boldsymbol{\tau}\cdot\hat{\boldsymbol{r}}\sin\Theta(r)\Big)\rho(\boldsymbol{x},\boldsymbol{x})$$
$$\text{(C.5)}$$

$$\sin\Theta(r) = \frac{m}{m_\pi^2 f_\pi^2}N_c\text{tr}\int\frac{d\Omega}{4\pi}\Big(\sin\Theta(r) - i\gamma_5\boldsymbol{\tau}\cdot\hat{\boldsymbol{r}}\cos\Theta(r)\Big)\rho(\boldsymbol{x},\boldsymbol{x}), \qquad \text{(C.6)}$$

$$\omega(r) = \frac{g_V^2}{4m_\rho^2}N_c\text{tr}\int\frac{d\Omega}{4\pi}\,b(\boldsymbol{x},\boldsymbol{x}), \qquad \text{(C.7)}$$

$$G(r) = -\frac{g_V^2}{4m_\rho^2}N_c\text{tr}\int\frac{d\Omega}{4\pi}\Big((\boldsymbol{\gamma}\times\hat{\boldsymbol{r}})\cdot\boldsymbol{\tau}\Big)\rho(\boldsymbol{x},\boldsymbol{x}), \qquad \text{(C.8)}$$

$$F(r) = -\frac{g_V^2}{4m_\rho^2}N_c\text{tr}\int\frac{d\Omega}{4\pi}\,\beta\Big(3(\boldsymbol{\sigma}\cdot\hat{\boldsymbol{r}})(\boldsymbol{\tau}\cdot\hat{\boldsymbol{r}}) - (\boldsymbol{\sigma}\cdot\boldsymbol{\tau})\Big)\rho(\boldsymbol{x},\boldsymbol{x}), \qquad \text{(C.9)}$$

$$H(r) = \frac{g_V^2}{4m_\rho^2}N_c\text{tr}\int\frac{d\Omega}{4\pi}\,\beta\Big((\boldsymbol{\sigma}\cdot\hat{\boldsymbol{r}})(\boldsymbol{\tau}\cdot\hat{\boldsymbol{r}}) - (\boldsymbol{\sigma}\cdot\boldsymbol{\tau})\Big)\rho(\boldsymbol{x},\boldsymbol{x}). \qquad \text{(C.10)}$$

The traces are over Dirac and isospin indices only. Note that only the ω meson profile is given by a trace over the baryon number density. All other meson profiles have as "source" the scalar quark density. Without regularization the difference between both densities would have only been a factor $\gamma_0 = \beta$. However, the regularization makes the relation between the two densities, and therefore the relation between the ω and the other meson profiles, highly non-linear.

References

[ABJ69]   S. Adler, Phys. Rev. **177** (1969) 2426; J. S. Bell and R. Jackiw, Nuov. Cim. **60A** (1969) 47.

[Al90]   R. Alkofer, Phys. Lett. **B236** (1990) 310.

[AHW95]   R. Alkofer, H. Reinhardt and H. Weigel, "Baryons as Chiral Solitons in the Nambu–Jona-Lasinio Model", Physics Reports, in press.

[ANW83]   G. S. Adkins, C. R. Nappi and E. Witten, Nucl. Phys. **B228** (1983) 552.

[AR91]   R. Alkofer and H. Reinhardt, Phys. Lett. **B244** (1991) 461.
             R. Alkofer, H. Reinhardt, H. Weigel and U. Zückert, Phys. Rev. Lett. **69** (1992) 1874; Phys. Lett. **B298** (1993) 132.

[AR92]   R. Alkofer and H. Reinhardt, Z. Phys. **A343** (1992) 79.

[AS65]   M. Abramowitz and I. A. Stegun, "Handbook of Mathematical Functions", Dover Publ., New York, 1965.

[AWZ94]   R. Alkofer and H. Weigel, Comput. Phys. Comm. **82** (1994) 30;
             U. Zückert, R. Alkofer and H. Weigel, Comput. Phys. Comm. **82** (1994) 42;
             H. Weigel and R. Alkofer, Comput. Phys. Comm. **82** (1994) 57.

[Ba92]   P. v. Baal, Nucl. Phys. **B369** (1992) 259.

[BAR92]   A. Buck, R. Alkofer and H. Reinhardt, Phys. Lett. **B286** (1992) 29.

[Be84]   M. V. Berry, Proc. R. Soc. London **A392** (1984) 45.

[BHS88]   A. H. Blin, B. Hiller and M. Schaden, Z. Phys. **A331** (1988) 75.

[BJM88]   V. Bernard, R. L. Jaffe and Ulf-G. Meißner, Nucl. Phys. **B308** (1988) 753
             V. Bernard, Ulf-G. Meißner, A. H. Blin and B. Hiller, Phys. Lett. **B253** (1991) 443

[BP89]   N. Brown and M. R. Pennigton, Phys. Rev. **D38** (1988) 2266; **D39** (1989) 2723.

[Co90]   S. Coleman, Aspects of Symmetry, Selected Erice Lectures, Cambridge University Press, 1990.

[CK85]   C. Callan and I. Klebanov, Nucl. Phys. **B262**, 365 (1985).

[CL88]   T. P. Cheng and L. F. Li, "Gauge Theory of Elementary Particles", Clardenon Press, 1988.

[CR85]   R. T. Cahill and C. D. Roberts, Phys. Rev. **D32** (1985) 2418.

[CRP87]   R. T. Cahill, C. D. Roberts and J. Praschifka, Phys. Rev. **D36** (1987) 2804.
             J. Praschifka, R. T. Cahill and C. D. Roberts, Int. J. Mod. Phys. **A4** (1989) 4929.
             R. T. Cahill, J. Praschifka and C. J. Burden, Australian Journal of Physics, **42** (1989) 161.

[ER86]   D. Ebert and H. Reinhardt, Nucl. Phys. **B271** (1986) 188.

[ERV94]   D. Ebert, H. Reinhardt and M. Volkov, "Effective Hadron Theory", in Progress in Particle and Nuclear Physics, vol. **33**, ed. A. Fäßler, Pergamon Press, Oxford, 1994.

[FP67]   L. D. Fadeev and V. N. Popov, Phys. Lett. **25B** (1967) 29.

[Fu80]   K. Fujikawa, Phys. Rev. **D21** (1980) 2848.

[FW71]   A.L. Fetter and J.D. Walecka, Quantum Theory of Many–Particle Sysytems, McGraw-Hill, New York, 1971.

[GMOR68]   M. Gell-Mann, R. J. Oakes and B. Renner, Phys. Rev. **175** (1968) 2195.
[Go59]   L. P. Gorkov, Sov. Phys. JETP **9** (1959) 1364; *see also sect. 38 of* A. A. Abrikosov, L. P. Gorkov and I. E. Dzyaloshinskii, "Methods of Quantum Field Theory in Statistical Physics", Prentice Hall Inc., Englewood Cliffs, N.J., 1963.
[Gr78]   V. N. Gribov, Nucl. Phys. **B139** (1978) 1.
[HK94]   T. Hatsuda and T. Kunihiro, Phys. Rep. **247** (1994) 221.
[Ho93]   Proc. Int. Workshop "Baryons as Skyrme Solitons", G. Holzwarth (Ed.), World Scientific P.C., Singapore 1993.
[IZ85]   C. Itzykson and J.-B. Zuber, "Quantum Field Theory", McGraw Hill, 1985.
[KLVW90]   S. Klimt, M. Lutz, U. Vogl and W. Weise, Nucl. Phys. **A516** (1990) 429.
[KR84]   S. Kahana and G. Ripka, Nucl. Phys. **A429** (1984) 462.
[LR94]   K. Langfeld and H. Reinhardt, Nucl. Phys. **A579** (1994) 472.
[LY62]   T. D. Lee and C. N. Yang, Phys. Rev. **128** (1962) 885.
[MGG89]   Th. Meißner, F. Grümmer and K. Goeke, Phys. Lett. **B227** (1989) 296.
   Th. Meißner and K. Goeke, Nucl. Phys. **A524** (1991) 719.
[MKW87]   Ulf.-G. Meißner, N. Kaiser and W. Weise, Nucl. Phys. **A466** (1987) 685;
   Ulf-G. Meißner, N. Kaiser, H. Weigel and J. Schechter, Phys. Rev. **D39** (1989) 1956.
[MR86]   N. S. Manton and P. J. Ruback, Phys. Lett. **B181** (1986) 137.
[MRWSGG92]   Th. Meißner, G. Ripka, R. Wünsch, P. Sieber, F. Grümmer and K. Goeke, Phys. Lett. **B299** (1993) 183.
[Na60]   Y. Nambu, Phys. Rev. **117** (1960) 648.
[NJL61]   Y. Nambu and G. Jona-Lasinio, Phys. Rev. **122** (1961) 345; **124** (1961) 246.
[PT84]   P. Pascual and R. Tarrach, "QCD: Renormalization for the Practitioner", Springer–Verlag, 1984.
[PW91]   N. W. Park and H. Weigel, Phys. Lett. **B268** (1991) 155; Nucl. Phys. **A541** (1992) 453.
[RA88]   H. Reinhardt and R. Alkofer, Phys. Lett. **B207** (1988) 482
   R. Alkofer and H. Reinhardt, Z. Phys. **C45** (1989) 245
[RD89]   H. Reinhardt and B. V. Dang, Nucl. Phys. **A500** (1989) 563.
[Re80]   H. Reinhardt, Nucl. Phys. **A346** (1980) 1.
[Re89]   H. Reinhardt, Nucl. Phys. **A503** (1989) 825.
[Re90]   H. Reinhardt, Phys. Lett. **B244** (1990) 316.
[Re90a]   H. Reinhardt, Phys. Lett. **B248** (1990) 365.
[Re91]   H. Reinhardt, Phys. Lett. **B257** (1991) 375.
[RS80]   P. Ring and P. Schuck, The Nuclear Many–Body Problem, Springer, New York, 1980.
[RW88]   H. Reinhardt and R. Wünsch, Phys. Lett. **B215** (1988) 577; **B 230** (1989) 93.
[RW94]   C. D. Roberts und A. G. Williams, Progress in Particle and Nuclear Physics, Vol. 33, p. 477, ed. A. Fäßler, Pergamon Press, Oxford, 1994.
[SAA91]   L. v. Smekal, P. A. Amundsen and R. Alkofer, Nucl. Phys. **A529** (1991) 663.

[SARW93]   J. Schlienz, R. Alkofer, H. Reinhardt and H. Weigel, Phys. Lett. **B315** (1993) 6.

[Sch51]   J. Schwinger, Phys. Rev. **82** (1951) 664.

[Sk61]   T. H. R. Skyrme, Proc. Roy. Soc. **A260** (1961) 127.

[Sm94]   L. v. Smekal, Ph. D. Thesis, Tübingen University 1994.

[SRAL90]   M. Schaden, H. Reinhardt, P. A. Amundsen and M. Lavelle, Nucl. Phys. **B339** (1990) 595.

[SW85]   B. D. Serot and J. D. Walecka, "The relativistic nuclear many-body problem", Advances in Nuclear Physics, eds. J. W. Negele and E. Vogt, 1985.

[tH74]   G. 't Hooft, Nucl. Phys. **B72** (1974) 461.

[tH76]   G. 't Hooft, Phys. Rev. **D14** (1976) 3432, **D18** (1978) 2199E.

[WAR92]   H. Weigel, R. Alkofer and H. Reinhardt, Nucl. Phys. **B387** (1992) 638.

[WAR94]   H. Weigel, R. Alkofer and H. Reinhardt, Nucl. Phys. **A576** (1994) 477.

[WAR95]   H. Weigel, R. Alkofer and H. Reinhardt, Nucl. Phys. **A582** (1995) 484.

[WAW93]   C. Weiss, R. Alkofer and H. Weigel, Mod. Phys. Lett. **A8** (1993) 79.

[WBAR93]   C. Weiss, A. Buck, R. Alkofer and H. Reinhardt, Phys. Lett. **B312** (1993) 6.

[WRA93]   H. Weigel, H. Reinhardt and R. Alkofer, Phys. Lett. **B313** (1993) 377.

[Wi79]   E. Witten, Nucl. Phys. **B160** (1979) 57.

[WSPM91]   H. Weigel, J. Schechter N. W. Park and Ulf-G. Meißner, Phys. Rev. **D43** (1991) 869.

[WT92]   T. Watabe and H. Toki, Prog. Theor. Phys. **87** (1992) 651;
            P. Sieber, Th. Meißner, F. Grümmer and K. Goeke, Nucl. Phys. **A547** (1992) 459.

[WZAR95]   H. Weigel, R. Alkofer and H. Reinhardt, Nucl. Phys. **A585** (1995) 513.

[YA88]   H. Yabu and K. Ando, Nucl. Phys. **B301** (1988) 601.

[ZARW]   U. Zückert, R. Alkofer, H. Reinhardt and H. Weigel, work in progress.

[ZARW94]   U. Zückert, R. Alkofer, H. Reinhardt and H. Weigel, Nucl. Phys. **A570** (1994) 445.

[ZARW95]   U. Zückert, R. Alkofer, H. Reinhardt and H. Weigel, Mod. Phys. Lett. **A10** (1995) 67.

[Zw92]   D. Zwanziger, Nucl. Phys. **B378** (1992) 525, Nucl. Phys. **B412** (1994) 657.

List of Figures

114

List of Tables

Springer-Verlag
and the Environment

We at Springer-Verlag firmly believe that an international science publisher has a special obligation to the environment, and our corporate policies consistently reflect this conviction.

We also expect our business partners – paper mills, printers, packaging manufacturers, etc. – to commit themselves to using environmentally friendly materials and production processes.

The paper in this book is made from low- or no-chlorine pulp and is acid free, in conformance with international standards for paper permanency.

Lecture Notes in Physics

For information about Vols. 1–419
please contact your bookseller or Springer-Verlag

New Series m: Monographs